T0299121

# LONDON MATHEMATICAL SOCIETY STUDENT TEXTS

Managing editor: Professor D. Benson,
Department of Mathematics, University of Aberdeen, UK

LONDON MATHEMATICAL SOCIETY STUDENT
TEXTS 73

# Groups, Graphs and Trees

## An Introduction to the Geometry of Infinite Groups

JOHN MEIER
*Lafayette College*

CAMBRIDGE
UNIVERSITY PRESS

CAMBRIDGE UNIVERSITY PRESS
Cambridge, New York, Melbourne, Madrid, Cape Town, Singapore,
São Paulo, Delhi, Dubai, Tokyo, Mexico City

Cambridge University Press
The Edinburgh Building, Cambridge CB2 8RU, UK

Published in the United States of America by Cambridge University Press, New York

www.cambridge.org
Information on this title: www.cambridge.org/9780521895453

First published 2008

*A catalogue record for this publication is available from the British Library*

*Library of Congress Cataloguing in Publication Data*

Meier, John, 1965-
Groups, graphs, and trees : an introduction to the geometry of infinite groups / John Meier.
p.   cm.
Includes bibliographical references and index.
ISBN 978-0-521-89545-3 (hardback) - ISBN 978-0-521-71977-3 (pbk.)
1. Infinite groups.   I. Title.
QA178.M45 2008
512'.2-dc22      2008012848

ISBN 978-0-521-89545-3 Hardback
ISBN 978-0-521-71977-3 Paperback

For my driver, Piotr

# Contents

# Preface

Groups are algebraic objects, consisting of a set with a binary operation that satisfies a short list of required properties: the binary operation must be associative; there is an identity element; and every element has an inverse. Presenting groups in this formal, abstract algebraic manner is both useful and powerful. Yet it avoids a wonderful geometric perspective on group theory that is also useful and powerful, particularly in the study of infinite groups. This perspective is hinted at in the combinatorial approach to finite groups that is often seen in a first course in abstract algebra. It is my intention to bring the geometric perspective forward, to establish some elementary results that indicate the utility of this perspective, and to highlight some interesting examples of particular infinite groups along the way. My own bias is that these groups are just as interesting as the theorems.

The topics covered in this book fit inside of "geometric group theory," a field that sits in the impressively large intersection of abstract algebra, geometry, topology, formal language theory, and many other fields. I hope that this book will provide an introduction to geometric group theory at a broadly accessible level, requiring nothing more than a single-semester exposure to groups and a naive familiarity with the combinatorial theory of graphs.

The chapters alternate between those devoted to general techniques and theorems (odd numbers) and brief chapters introducing some of the standard examples of infinite groups (even numbers). Chapter 2 presents a few groups generated by reflections; Chapter 4 presents the Baumslag–Solitar group $BS(1, 2)$ in terms of linear functions; Chapter 6 is the Gupta–Sidki variant of Grigorchuk's group; the Lamplighter group is discussed in Chapter 8; and Thompson's group $F$ is the subject of Chapter 10. When I taught this material at Lafayette College I referred

to the material in these even numbered chapters as "field trips to the
Zoo of Infinite Groups."

The first chapter should be relatively easy to work through, as it
reviews material on groups (mainly finite groups), group actions, and
the combinatorial theory of graphs. It establishes quite a bit of notation
and introduces the construction of Cayley graphs. While some material
in this chapter may be new to the reader, most of it should seem to be
a repackaging of ideas that she or he has previously encountered.

Chapter 3 is an introduction to free groups and free products of
groups. Chapters 5 and 7 are devoted to connections between finitely
generated groups and formal language theory. Chapters 9 and 11 deal
with the geometry of infinite groups, with Chapter 9 focusing on what
might be called the "fine geometry" of Cayley graphs, while Chapter 11
treats what is called the large-scale geometry of groups.

While no background beyond elementary group theory is necessary for
this book, a broader undergraduate exposure to mathematics is certainly
helpful. My experience in the classroom indicates that the material in
Chapter 7 is demanding for people who have not previously encountered
formal language theory. Similarly Chapter 11 is easier for people who
have had a course in real analysis. Because hyperbolic geometry is not
a standard undergraduate topic, Gromov's theory of hyperbolic groups
does not appear in this book. Similarly, because algebraic topology is
not a standard undergraduate topic, I have avoided fundamental groups
and covering spaces.

There are two forms of exercises in this book. A few exercises are
embedded within the chapters. These should be done, at least at the
level of the reader convincing themselves they know how to do them,
while reading through the material. There are also end-of-chapter exercises that are arranged in the order that material is presented in the
chapter. Some of these end-of-chapter exercises are challenging but most
are reasonably accessible.

*Groups, Graphs and Trees* was developed from notes used in two undergraduate course offerings at Lafayette College, and it can certainly serve
as a primary text for an advanced undergraduate course. It should also
be useful as a text for a reading course and as a gentle introduction to
geometric group theory for mathematicians with a broader background
than this. An undergraduate course that attempted to cover this text,
omitting no details, in one semester, would have to move at a rather
brisk pace. The critical background information is contained in the first
five chapters and those should not be trimmed. With a bit of forethought

an instructor can cover much of the rest of this book, if for example the material in Chapter 7 or Chapter 11 is presented more as a colloquium than as course material. My own hope is that various classes will find the space in their semester to pursue tangents of interest to them, and then let me know the results of their exploration.

I have many people to thank. My wife Trisha and son Robert were unreasonably supportive of this project. Many students provided important feedback as I fumbled through the process of presenting this bit of advanced mathematics at an elementary level: George Armagh, Kari Barkley, Jenna Bratz, Jacob Carson, Joellen Cope, Joe Dudek, Josh Goldstein, Ekaterina Jager, Brian Kronenthal, Rob McEwen, and Zachary Reiter. I also benefited from extensive feedback given by my colleagues Ethan Berkove and Jon McCammond. Finally, a number of anonymous referees provided comments on various draft chapters. I was impressed by the fact that there was no intersection between the comments provided by students, the comments provided by colleagues, and the anonymous referees!

# 1
# Cayley's Theorems

As for everything else, so for a mathematical theory: beauty can be perceived but not explained.

−Arthur Cayley

An introduction to group theory often begins with a number of examples of finite groups (symmetric, alternating, dihedral, ...) and constructions for combining groups into larger groups (direct products, for example). Then one encounters Cayley's Theorem, claiming that every finite group can be viewed as a subgroup of a symmetric group. This chapter begins by recalling Cayley's Theorem, then establishes notation, terminology, and background material, and concludes with the construction and elementary exploration of Cayley graphs. This is the foundation we use throughout the rest of the text where we present a series of variations on Cayley's original insight that are particularly appropriate for the study of infinite groups.

Relative to the rest of the text, this chapter is gentle, and should contain material that is somewhat familiar to the reader. A reader who has not previously studied groups and encountered graphs will find the treatment presented here "brisk."

## 1.1 Cayley's Basic Theorem

You probably already have good intuition for what it means for a group to act on a set or geometric object. For example:

- The cyclic group of order $n$ – denoted $\mathbb{Z}_n$ – acts by rotations on a regular $n$-sided polygon.

1

- The dihedral group of order $2n$ – denoted $D_n$ – also acts on the regular $n$-sided polygon, where the elements either rotate or reflect the polygon.
- We use $\text{SYM}_n$ to denote the symmetric group of all permutations of $[n] = \{1, 2, \ldots, n\}$. (More common notations are $S_n$ and $\Sigma_n$.) By its definition, $\text{SYM}_n$ acts on this set of numbers, as does its index 2 subgroup, the alternating group $A_n$, consisting of the even permutations.
- Matrix groups, such $\text{GL}_n(\mathbb{R})$ (the group of invertible $n$-by-$n$ matrices with real number entries), act on vector spaces.

Because the general theme of this book is to study groups via actions, we need a bit of notation and a formal definition.

**Convention 1.1.** If $X$ is a mathematical object (such as a regular polygon or a set of numbers), then we use $\text{SYM}(X)$ to denote all bijections from $X$ to $X$ that preserve the indicated mathematical structure. For example, if $X$ is a set, then $\text{SYM}(X)$ is simply the group of permutations of the elements of $X$. In fact, if $n = |X|$ then $\text{SYM}(X) \approx \text{SYM}_n$. Moreover, if $X$ and $X'$ have the same cardinality, then $\text{SYM}(X) \approx \text{SYM}(X')$. If $X$ is a regular polygon, then angles and lengths are important, and $\text{SYM}(X)$ will be composed of rotations and reflections (and it will in fact be a dihedral group). Similarly, if $X$ is a vector space, then $\text{SYM}(X)$ will consist of bijective linear transformations.

What we are referring to as "$\text{SYM}(X)$" does have a number of different names in different contexts within mathematics. For example, if $G$ is a group, then the collection of its symmetries is referred to as $\text{AUT}(G)$, the group of automorphisms. If we are working with the Euclidean plane, $\mathbb{R}^2$, and are considering functions that preserve the distance between points, then we are looking at $\text{ISOM}(\mathbb{R}^2)$, the group of isometries of the plane.

It is quite useful to have individual names for these groups, as their names highlight what mathematical structures are being preserved. Our convention of lumping these various groups all together under the name "$\text{SYM}$" is vague, but we believe that in context it will be clear what is intended, and we like the fact that this uniform terminology emphasizes that these various situations where groups arise are not all that different.[1] One egregious example, which highlights the need to be care-

---

[1] In his book, *Symmetry*, Hermann Weyl wrote: "[W]hat has indeed become a guiding principle in modern mathematics is this lesson: *Whenever you have to deal with a structure-endowed entity $\Sigma$ try to determine its group of automorphisms,* the group of those element-wise transformations which leave all structural relations

ful in using our convention, comes from the integers. If the integers are thought of as simply a set, containing infinitely many elements, then $\text{Sym}(\mathbb{Z})$ is an infinite permutation group, which contains $\text{Sym}_n$ for any $n$. On the other hand, if $\mathbb{Z}$ denotes the group of integers under addition, then $\text{Sym}(\mathbb{Z}) \approx \mathbb{Z}_2$. (The only non-trivial automorphism of the group of integers sends $n$ to $-n$ for all $n \in \mathbb{Z}$.)

**Definition 1.2.** An *action* of a group $G$ on a mathematical object $X$ is a group homomorphism from $G$ to $\text{Sym}(X)$. Equivalently, it is a map from $G \times X \to X$ such that

1. $e \cdot x = x$, for all $x \in X$; and
2. $(gh) \cdot x = g \cdot (h \cdot x)$, for all $g, h \in G$ and $x \in X$.

We denote "$G$ acts on $X$" by $G \curvearrowright X$.

If one has a group action $G \curvearrowright X$, then the associated homomorphism is a *representation* of $G$. The representation is *faithful* if the map is injective. In other words, it is faithful if, given any non-identity element $g \in G$, there is some $x \in X$ such that $g \cdot x \neq x$.

**Example 1.3.** The dihedral group $D_n$ is the symmetry group of a regular $n$-gon. As such, it also permutes the vertices of the $n$-gon, hence there is a representation $D_n \to \text{Sym}_n$. As every non-identity element of $D_n$ moves at least $(n-2)$ vertices, this representation is faithful.

**Remark 1.4** (left vs. right). In terms of avoiding confusion, this is perhaps the most important remark in this book. Because not all groups are abelian, it is very important to keep left and right straight. All of our actions will be *left* actions (as described above). We have chosen to work with left actions since it matches function notation and because left actions are standard in geometric group theory and topology.

Groups arise in a number of different contexts, most commonly as symmetries of any one of a number of possible mathematical objects $X$. In these situations, one can often understand the group directly from our understanding of $X$. The dihedral and symmetric groups are two examples of this. However, groups are abstract objects, being merely a set with a binary operation that satisfies a certain minimal list of requirements. Cayley's Theorem shows that the abstract notion of a group and the notion of a group of permutations are one and the same.

undisturbed. You can expect to gain a deep insight into the constitution of $\Sigma$ in this way." Our use of $\text{Sym}(\Sigma)$ instead of $\text{Aut}(\Sigma)$ is a small notational deviation from Weyl's recommendation.

**Theorem 1.5** (Cayley's Basic Theorem). *Every group can be faithfully represented as a group of permutations.*

*Proof.* The objects that $G$ permutes are the elements of $G$. In this proof we use "$\textsc{Sym}_G$" to denote $\textsc{Sym}(G)$, to emphasize that "$G$" denotes the underlying *set* of elements, not the group. The permutation associated to $g \in G$ is defined by left multiplication by $g$. That is, $g \mapsto \pi_g \in \textsc{Sym}_G$ where $\pi_g(h) = g \cdot h$ for all $h \in G$. This is a permutation of the elements of $G$, since if $g \cdot h = g \cdot h'$, then by left cancellation, $h = h'$. Denote the map taking the element $g$ to the permutation $\pi_g$ by $\pi : G \to \textsc{Sym}_G$.

To check that $\pi$ is a group homomorphism we need to verify that $\pi(gh) = \pi(g) \cdot \pi(h)$. In other words, we need to show that $\pi_{gh} = \pi_g \cdot \pi_h$. We do this by evaluating what each side does to an arbitrary element of $G$. We denote the arbitrary element by "$x$", thinking of it as a variable. The permutation $\pi_{gh}$ takes $x \mapsto (gh) \cdot x$, and successively applying $\pi_h$ then $\pi_g$ sends $x \mapsto h \cdot x \mapsto g \cdot (h \cdot x)$. Thus checking that $\phi$ is a homomorphism amounts to verifying the associative law: $(gh) \cdot x = g \cdot (h \cdot x)$. As this is part of the definition of a group, the equation holds.

In order to see that the map is faithful it suffices to show that no non-identity element is mapped to the trivial permutation. One can do this by simply noting that if $g \in G \setminus \{e\}$, then $g \cdot e = g$, hence $\pi_g(e) = g$, and so $\pi_g$ is not the identity (or trivial) permutation. $\square$

The proof of Cayley's Basic Theorem constructs a representation of $G$ as a group of permutations of itself. Before moving on we should examine what these permutations look like in some concrete situations. We first consider $\textsc{Sym}_3$, the group of all permutations of three objects.

**Notation 1.6** (cycle notation). In describing elements of $\textsc{Sym}_n$ we use cycle notation, and multiply (that is, compose permutations) right to left. This matches with our intuition from functions where $f \circ g(x)$ means that you first apply $g$ then apply $f$, and it is consistent with our use of left actions. Here is a concrete example: $(12)(35) \in \textsc{Sym}_5$ is the element that transposes 1 and 2, as well as 3 and 5; the element $(234)$ sends 2 to 3, 3 to 4 and 4 to 2; the product $(12)(35) \cdot (234) = (12534)$. (The product is not $(13542)$, which is the result of multiplying left to right.)

**Example 1.7.** The group SYM₃ has six elements, shown as disjoint vertices in Figure 1.1. The permutations described by Cayley's Basic Theorem – for the elements (12) and (123) – are also shown.

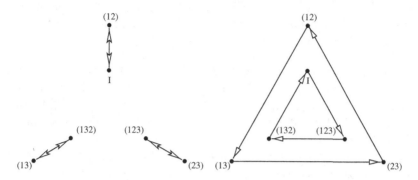

Fig. 1.1. The permutation of SYM₃ induced by (12) is shown on the left, and the permutation induced by (123) is shown on the right.

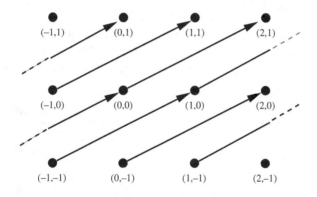

Fig. 1.2. The action of $(2, 1)$ on $\mathbb{Z} \oplus \mathbb{Z}$.

**Example 1.8.** In most introductions to group theory, Cayley's Basic Theorem is stated for *finite* groups. But we made no such assumption in our statement and the same proof as is given for finite groups works for infinite groups. Consider for example the direct product of two copies of the group of integers, $G = \mathbb{Z} \oplus \mathbb{Z}$. Here elements are represented by pairs of integers, and the binary operation is coordinatewise addition: $(a, b) + (c, d) = (a + c, b + d)$. In Figure 1.2 we have arranged the vertices corresponding to elements of $G$ as the integral lattice in the plane. The arrows indicate the permutation of the elements of $\mathbb{Z} \oplus \mathbb{Z}$ induced by the element $(2, 1)$.

## 1.2 Graphs

One of the key insights into the study of groups is that they can be viewed as symmetry groups of graphs. We refer to this as "Cayley's Better Theorem," which we prove in Section 1.5.2. In this section we establish some terminology from graph theory, and in the following section we discuss groups acting on graphs.

**Definition 1.9.** A *graph* $\Gamma$ consists of a set $V(\Gamma)$ of *vertices* and a set $E(\Gamma)$ of *edges*, each edge being associated to an unordered pair of vertices by a function "ENDS": $\text{ENDS}(e) = \{v, w\}$ where $v, w \in V$. In this case we call $v$ and $w$ the *ends* of the edge $e$ and we also say $v$ and $w$ are *adjacent*.

We allow the possibility that there are multiple edges with the same associated pair of vertices. Thus for two distinct edges $e$ and $e'$ it can be the case that $\text{ENDS}(e) = \text{ENDS}(e')$. We also allow loops, that is, edges whose associated vertices are the same. Graphs without loops or multiple edges are *simple* graphs.

Graphs are often visualized by making the vertices points on paper and edges arcs connecting the appropriate vertices. Two simple graphs are shown in Figure 1.3; a graph which is not simple is shown in Figure 1.4.

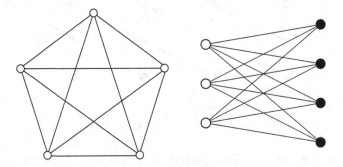

Fig. 1.3. The complete graph on five vertices, $K_5$, and the complete bipartite graph $K_{3,4}$.

There are a number of families of graphs that arise in mathematics. The *complete graph* on $n$ vertices has exactly one edge joining each pair of distinct vertices, and is denoted $K_n$. At the opposite extreme are the *null graphs*, which have no edges.

A graph is *bipartite* if its vertices can be partitioned into two subsets – by convention these subsets are referred to as the "black" and "white" vertices – such that, for every $e \in E(\Gamma)$, ENDS$(e)$ contains one black vertex and one white vertex. The *complete bipartite* graphs are simple graphs whose vertex sets have been partitioned into two collections, $V_{\circ}$ and $V_{\bullet}$, with edges joining each vertex in $V_{\circ}$ with each vertex in $V_{\bullet}$. If $|V_{\circ}| = n$ and $|V_{\bullet}| = m$ then the corresponding complete bipartite graph is denoted $K_{n,m}$.

The *valence* or *degree* of a vertex is the number of edges that contain it. For example, the valence of any vertex in $K_n$ is $n - 1$. If a vertex $v$ is the vertex for a loop, that is an edge $e$ where ENDS$(e) = \{v, v\}$, then this loop contributes twice to the computation of the valence of $v$. For example, the valence of the leftmost vertex in the graph shown in Figure 1.4 is six.

A graph is *locally finite* if each vertex is contained in a finite number of edges, that is, if the valence of every vertex is finite.

An *edge path*, or more simply a *path*, in a graph consists of an alternating sequence of vertices and edges, $\{v_0, e_1, v_1, \ldots, v_{n-1}, e_n, v_n\}$ where ENDS$(e_i) = \{v_{i-1}, v_i\}$ (for each $i$). A graph is *connected* if any two vertices can be joined by an edge path. In Figure 1.4 we have indicated an

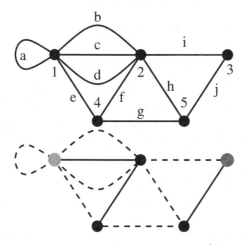

Fig. 1.4. On top is a graph which is not simple, with its vertices labelled by numbers and its edges labelled by letters. Below is the set of edges traversed in an edge path, joining the vertex labelled 1 to the vertex labelled 3, is indicated.

edge path from the leftmost vertex to the rightmost vertex. If $v_i$ is the vertex labelled $i$ and $e_\alpha$ is the edge labelled $\alpha$, then this path is:

$$\{v_1, e_a, v_1, e_e, v_4, e_g, v_5, e_h, v_2, e_b, v_1, e_d, v_2, e_i, v_3\}$$

This notation is obviously a bit cumbersome and is only feasible when
the graphs are small and the paths are short. If the graph is a simple
graph, then one really only needs to list the sequence of vertices, which
is a small economy of notation. In general, we will not need this level of
specificity in dealing with edge paths in graphs.

A *backtrack* is a path of the form $\{v, e, w, e, v\}$ where one has trav-
elled along the edge $e$ and then immediately returned along $e$. A path
is *reduced* if it contains no backtracks.

A *cycle* or *circuit* is a non-trivial edge path whose first and last vertices
are the same, but no other vertex is repeated. The following paths in
the graph shown in Figure 1.4 are all cycles:

1.  $\{v_1, e_e, v_4, e_f, v_2, e_c, v_1\}$,
2.  $\{v_1, e_b, v_2, e_d, v_1\}$,
3.  $\{v_1, e_a, v_1\}$.

In Chapter 3 we study various groups that are closely connected to
trees. A *tree* is a connected graph with no cycles. If you have not encountered trees in a previous course, working through the following exercise
will help you gain some intuition for trees.

**Exercise 1.10.** Prove that the following conditions on a connected
graph $\Gamma$ are equivalent.

1.  $\Gamma$ is a tree.
2.  Given any two vertices $v$ and $w$ in $\Gamma$, there is a unique reduced
    edge path from $v$ to $w$.
3.  For every edge $e \in E(\Gamma)$, removing $e$ from $\Gamma$ disconnects the
    graph. (Note: Removing $e$ does not remove its associated ver-
    tices.)
4.  *If $\Gamma$ is finite then* $\#V(\Gamma) = \#E(\Gamma) + 1$.

While there are a number of interesting results about finite trees, in
this book we shall be mainly interested in infinite trees. In particular,
in Chapter 3 we explore groups that act on certain infinite, symmetric
trees.

**Definition 1.11** (regular and biregular trees). A *regular m-tree* is a tree
where every vertex has fixed valence $m$. For a given value of $m$ there is
only one regular $m$-tree, which we denote $\mathcal{T}_m$. Notice that, since every
vertex has valence $m$, the tree $\mathcal{T}_m$ is infinite when $m \geq 2$.

A graph is *biregular* if it is bipartite, and all the vertices in one class
have fixed valence $m$ and all the vertices in the other class have fixed

valence $n$. Thus, for example, the complete bipartite graphs are all biregular. Given the valences, there is a unique biregular tree, which we denote by $\mathcal{T}_{m,n}$. You can see an example in Figure 1.5.

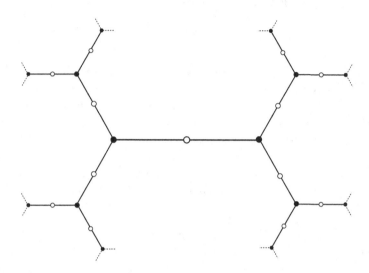

Fig. 1.5. The biregular tree $\mathcal{T}_{2,3}$.

It is often convenient to think of graphs as geometric objects where each edge is identified with the unit interval $[0,1]$, where 0 corresponds to one of the associated vertices and 1 to the other. This convention allows us to refer to *midpoints* of edges, for example. The graphs shown in Figures 1.3, 1.4, and 1.5 have distorted this metric, meaning that if you measured the lengths of the edges in a given figure, those lengths will not all be the same. (This will occur in almost all of the graphs drawn in this book.) In addition to allowing us to specify points on edges, this geometric perspective allows us to think of paths as parametrized curves. Our convention will be that, given an edge path

$$\omega = \{v_0, e_1, v_1, e_2, \ldots, e_n, v_n\}$$

in a graph $\Gamma$, there is an associated function $p_\omega : [0,1] \to \Gamma$ where $p_\omega(i/n) = v_i$ and $p_\omega$ is linear when restricted to each of the subintervals $[i/n, (i+1)/n]$.

**Remark 1.12.** Our view of graphs is more topological than combinatorial. The reader who is a bit uneasy about our thinking of graphs as geometric objects, where edges have lengths, might want to look up the definition of CW complexes, as we are viewing graphs as 1-dimensional CW complexes. We have not introduced this terminology or explicitly used the associated definition as it requires an understanding of topological spaces and quotient topologies.

There is one final variation on graphs that we will encounter in this text:

**Definition 1.13.** A *directed* graph consists of a vertex set $V$ and an edge set $E$ of ordered pairs of vertices. Thus each edge has an *initial* vertex and a *terminal* vertex. Graphically this direction is often indicated via an arrow on the edge. In thinking of directed graphs geometrically we assume that the initial vertex is identified with $0 \in [0,1]$ and the terminal vertex with $1 \in [0,1]$.

If we say a directed graph is connected we mean the underlying undirected graph is connected. (One can study "directed-connectedness" but that will not be relevant for us.)

In addition to directions on the edges, there are other sorts of decorations one can add to a graph. For example, one can have a set of labels $\mathcal{L}$ and a function $\ell_V : V(\Gamma) \to \mathcal{L}$ that provides a labelling of the vertices. Or one could label the edges via $\ell_E : E(\Gamma) \to \mathcal{L}$, where the set of labels might be the same or different than the labels for the vertices. As an example, in Figure 1.4 we have shown a labelled graph where the vertices have been labelled with numbers and the edges with lower case letters.

## 1.3 Symmetry Groups of Graphs

Many important finite groups arise as symmetry groups of geometric objects. The dihedral groups are the symmetry groups of regular $n$-gons; the symmetric group $\mathrm{SYM}_n$ is isomorphic to the symmetry group of the regular $(n-1)$-dimensional simplex, for example the convex hull of

$$\{(1,0,0,\ldots,0),(0,1,0,\ldots,0),\ldots,(0,\ldots,0,1)\} \subset \mathbb{R}^n \ ;$$

the alternating subgroup $A_n$ is isomorphic to the subgroup of symmetries of the regular $(n-1)$-dimensional simplex consisting of rotations in $\mathbb{R}^n$.

In this section we explore a similar theme, namely, we explore symmetry groups of graphs.

**Definition 1.14.** A *symmetry* of a graph $\Gamma$ is a bijection $\alpha$ taking vertices to vertices and edges to edges such that if $\mathrm{ENDS}(e) = \{v, w\}$, then

ENDS($\alpha(e)$) = $\{\alpha(v), \alpha(w)\}$. The *symmetry group of* $\Gamma$ is the collection of all its symmetries. We denote this group by SYM($\Gamma$).[2]

**Example 1.15.** The symmetry group of the complete graph $K_n$ is isomorphic to SYM$_n$.

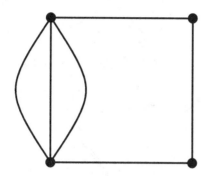

Fig. 1.6. The graph for Example 1.16.

**Example 1.16.** The symmetry group of the graph $\Gamma$ shown in Figure 1.6 is isomorphic to SYM$_3 \oplus \mathbb{Z}_2$ (or SYM$_3 \oplus$ SYM$_2$ if you prefer the symmetry in notation). The symmetries come from permuting the multiple edges joining the leftmost vertices of $\Gamma$ (which gives the SYM$_3$ factor) and from a reflection through a horizontal axis (which gives the SYM$_2$ factor). The fact that these two types of symmetries commute with each other leads to the direct product structure.

**Definition 1.17.** Let $G$ be a subgroup of SYM($\Gamma$). Then $G$ is *vertex transitive* if, given any two vertices, $v$ and $v'$, there is an $\alpha \in G$ where $\alpha(v) = v'$. The symmetry groups of the complete graphs are all vertex transitive, while the symmetry group of a complete bipartite graph $K_{n,m}$ is not vertex transitive when $n \neq m$. (Why?)

A group $G < $ SYM($\Gamma$) is *edge transitive* if, given any two edges, $e$ and $e'$, there is a symmetry $\alpha \in G$ where $\alpha(e) = e'$.

A *flag* consists of a pair $(v, e)$, where $v \in$ ENDS($e$). The group $G$ is *flag transitive* if, given any two flags, $(v, e)$ and $(v', e')$, there is a symmetry $\alpha \in G$ where $\alpha(v) = v'$ and $\alpha(e) = e'$.

The symmetry group $G$ acts *simply transitively* on the vertices (edges, flags, etc.) if it is vertex transitive, and given any two vertices $v$ and $v'$

---

[2] Graph theorists usually refer to *automorphisms* of a graph, and call the collection of all automorphisms the *automorphism group of the graph*.

there is a unique $\alpha \in G$ such that $\alpha(v) = v'$ (respectively, $\alpha(e) = e'$, etc.)

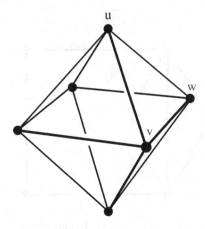

Fig. 1.7. The graph for Example 1.18.

**Example 1.18.** Let $\Gamma$ be the vertices and edges of an octahedron, as in Figure 1.7. The vertices of $\Gamma$ can be partitioned into three pairs of opposite vertices. These pairs are: the top–bottom pair (containing the vertex labelled $u$ in Figure 1.7); the front–back pair (containing $v$); and the left–right pair (containing $w$). Using reflections one can exchange the two vertices in the top–bottom, front–back, or left–right pair, leaving the other four vertices fixed. Direct inspection shows that any two such reflections commute, so

$$(\mathbb{Z}_2)^3 = \mathbb{Z}_2 \oplus \mathbb{Z}_2 \oplus \mathbb{Z}_2 < \text{Sym}(\Gamma).$$

The action of this subgroup is not transitive on the vertices of $\Gamma$. However, given any two sets of opposing vertices, there is a reflection that exchanges those two sets, which fixes the remaining pair of vertices. These reflections generate a subgroup that permutes the pairs of opposite vertices, giving a subgroup isomorphic to $\text{Sym}_3$. Let the subgroup of $\text{Sym}(\Gamma)$ generated by this copy of $\text{Sym}_3$ and the copy of $(\mathbb{Z}_2)^3$ be denoted $H$. ($H$ is the collection of all possible products of elements of $\text{Sym}_3$ and $(\mathbb{Z}_2)^3$.) This subgroup is vertex transitive. To move, for example, the vertex labelled $u$, to vertex $v$, first exchange the top–bottom pair of vertices with the front–back pair of vertices. Then, if necessary, one can exchange the front–back pair in order to have the image of $u$ be $v$.

The subgroup $H$ described above is in fact the entire group of symmetries. To show this, let $\alpha \in \text{SYM}(\Gamma)$ be an arbitrary symmetry. Once we know where $\alpha$ moves the vertices $u, v$, and $w$, we have determined $\alpha$. Since the action of $H$ is vertex transitive, there is an $a \in H$ that takes $u$ to $\alpha(u)$. Thus $a^{-1}\alpha$ is a symmetry that fixes $u$. If $a^{-1}\alpha$ exchanges the front–back pair of vertices with the left–right pair of vertices, then there is a $b \in H$ that fixes the top–bottom pair of vertices and also exchanges the front–back pair of vertices with the left–right pair of vertices. Taking $b$ to be the identity if $a^{-1}\alpha$ did not make this exchange, one sees that $b^{-1}a^{-1}\alpha$ fixes $u$ and takes the front–back pair of vertices to itself, and similarly with the left–right pair. Thus there is a $c \in (\mathbb{Z}_2)^3 < H$ such that $c^{-1}b^{-1}a^{-1}\alpha$ fixes $u, v$, and $w$. But then $c^{-1}b^{-1}a^{-1}\alpha = e$, hence $\alpha = abc \in H$.[3]

Once one has this description of $\text{SYM}(\Gamma)$, it is not difficult to argue that not only is $\text{SYM}(\Gamma)$ vertex transitive, it is flag transitive.

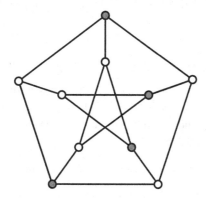

Fig. 1.8. This is the Petersen graph. A 4-vertex subgraph, a null graph, has been highlighted.

**Example 1.19.** Let $\Gamma$ be the Petersen graph, shown in Figure 1.8. Exploiting the symmetry present in this particular picture, one sees that $D_5 < \text{SYM}(\Gamma)$. There is, however, much more symmetry in this graph than is immediately apparent from the figure.

---

[3] If you have already encountered semi-direct products of groups, then you should be able to show $\text{SYM}(\Gamma) \approx (\mathbb{Z}_2)^3 \rtimes \text{SYM}3$. We define semi-direct products in Chapter 8.

**Exercise 1.20.** The following questions explore the symmetries of the Petersen graph.

   a. Prove that the symmetry group of $\Gamma$ is vertex transitive by finding a symmetry that exchanges the set of "inner" vertices with the "outer" vertices.
   b. Prove that $\text{SYM}(\Gamma)$ is edge transitive as well. Is it also flag transitive?
   c. A 4-vertex subgraph, a null graph, is highlighted in Figure 1.8. There are five such subgraphs. Find them all. (Hint: Each one is determined by picking a pair of non-adjacent vertices on the "outer" ring.)
   d. Use the null graphs to construct a homomorphism

$$\phi : \text{SYM}(\Gamma) \to \text{SYM}_5$$

   e. Show that $\phi$ is onto.
   f. Show that the kernel of $\phi$ is trivial, hence, by the First Isomorphism Theorem, $\text{SYM}(\Gamma) \approx \text{SYM}_5$.

We now turn to a discussion of particular subgroups of symmetry groups of graphs.

**Lemma 1.21.** *If $\Gamma$ is a directed graph, the collection of all symmetries of $\Gamma$ that preserve every edge's direction forms a subgroup of $\text{SYM}(\Gamma)$.*

*Proof.* In order to prove a subset of a group is a subgroup, one only needs to show that the subset is closed under products and inverses. It is clear that if $g$ and $h$ preserve edge directions then so does their product and so do their inverses.    □

Essentially the same comments establish the following result.

**Lemma 1.22.** *If the edges and/or vertices of $\Gamma$ are labelled, the collection of all symmetries that preserve the edge and/or vertex labels forms a subgroup of $\text{SYM}(\Gamma)$.*

**Definition 1.23.** If $\Gamma$ comes with certain declared decorations – such as directed edges, labelled or colored vertices and/or edges, etc. – let $\text{SYM}^+(\Gamma)$ denote the subgroup of $\text{SYM}(\Gamma)$ that preserves all of the declared decorations.

**Exercise 1.24.** Consider the collection of complete bipartite graphs, $K_{n,m}$. Let $\text{SYM}^+(K_{n,m})$ be the group of symmetries of $K_{n,m}$ that preserve the bipartite structure. That is, if $g \in \text{SYM}^+(K_{n,m})$, then $g$ takes

white vertices to white vertices and black vertices to black vertices. (This is described more symbolically by writing $g \cdot V_\circ = V_\circ$ and $g \cdot V_\bullet = V_\bullet$.) Prove that $\text{SYM}^+(K_{n,m}) = \text{SYM}_n \oplus \text{SYM}_m$. What can you say about the full symmetry group $\text{SYM}(K_{n,m})$?
(Hint: It matters whether or not $n = m$.)

## 1.4 Orbits and Stabilizers

In this section we introduce terminology that gets used quite frequently when studying group actions.

**Definition 1.25.** Let $X$ be some mathematical object (perhaps $X$ is a set, a graph, a regular polygon, ...), and let $G$ act on $X$. If $x \in X$ then the *stabilizer* of $x$ is

$$\text{Stab}(x) = \{g \in G \mid g \cdot x = x\}.$$

Since composing two symmetries that fix $x$ yields a symmetry that still fixes $x$, and similarly the inverse of a symmetry fixing $x$ must fix $x$, we have:

**Lemma 1.26.** *For any $G \curvearrowright X$, and any $x \in X$, $\text{Stab}(x)$ is a subgroup of $G$.*

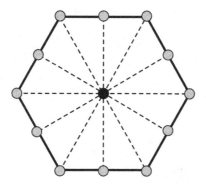

Fig. 1.9. The dihedral group $D_6$ acts on the regular hexagon. The stabilizer of any grey vertex is isomorphic to $\mathbb{Z}_2$. The stabilizer of the center point is all of $D_6$. Any point which is not indicated and which is not on one of the dotted lines, has a trivial stabilizer.

**Example 1.27.** The dihedral group $D_n$ acts on a regular $n$-gon. If $x$ is a vertex of this $n$-gon, or the exact middle of any edge, then $\text{Stab}(x) \approx \mathbb{Z}_2$.

Notice, however, that if $x$ and $x'$ are two vertices that are not directly opposite each other, then $\mathrm{Stab}(x) \neq \mathrm{Stab}(x')$. If $x$ is the center of the $n$-gon then $\mathrm{Stab}(x) \approx D_n$. If $x$ is any point that is not a vertex, the center of an edge, or on a line connecting the center of the polygon to a vertex or the center of an edge, then $\mathrm{Stab}(x) = \{e\}$.

**Example 1.28.** Let $\mathrm{SYM}_n \curvearrowright [n] = \{1, 2, \ldots, n\}$. Then $\mathrm{Stab}(i) \approx \mathrm{SYM}_{n-1}$ for any $i \in [n]$.

**Definition 1.29.** Let $G \curvearrowright X$. If for some $x \in X$ $\mathrm{Stab}(x) = \{e\}$, then $x$ is said to be *moved freely* by the action of $G$. The action $G \curvearrowright X$ is *free* if for every $x \in X$, $\mathrm{Stab}(x) = \{e\}$.

**Example 1.30.** Let $R$ be the subgroup of rotations inside of $D_n$. If $\mathcal{P}$ is a regular $n$-gon, let $\partial \mathcal{P}$ be its boundary, consisting of $n$ vertices and $n$ edges. The action of $R$ on $\partial \mathcal{P}$ is free, while the action of $R$ on $\mathcal{P}$ is not free.

**Definition 1.31.** Let $G \curvearrowright X$ and let $x \in X$. Then the *orbit* of $x$ is

$$\mathrm{Orb}(x) = \{g \cdot x \mid g \in G\}.$$

**Example 1.32.** The orbit of any vertex under the action of $D_n$ on a regular $n$-gon $\mathcal{P}$ is the set of all vertices. Similarly if $x$ is the midpoint of any edge in $\mathcal{P}$ then the orbit of $x$ is the set of all midpoints of edges.

Fix an action of a group $G$ on some $X$, and a particular $x \in X$. Then $g \cdot x = h \cdot x$ if and only if $g^{-1}h \cdot x = x$, that is, if and only if $g^{-1}h \in \mathrm{Stab}(x)$. This implies that $g \cdot x = h \cdot x$ if and only if the left cosets $g \cdot \mathrm{Stab}(x)$ and $h \cdot \mathrm{Stab}(x)$ are the same. This short argument establishes:

**Theorem 1.33.** *Given an action $G \curvearrowright X$ and any $x \in X$, there is a bijective correspondence between the set $\mathrm{Orb}(x)$ and left cosets of $\mathrm{Stab}(x)$, given by $g \cdot x \leftrightarrow g \cdot \mathrm{Stab}(x)$.*

A corollary of this result, which is often useful in computing the orders of finite groups, is the Orbit-Stabilizer Theorem:

**Corollary 1.34** (Orbit-Stabilizer Theorem). *Let $G$ be a finite group acting on $X$. Then, for any $x \in X$,*

$$|G| = |\mathrm{Stab}(x)| \cdot |\mathrm{Orb}(x)| \, .$$

*Proof.* The order of $G$ is the product of the order of $\text{Stab}(x)$ and the index of $\text{Stab}(x)$ in $G$. But the index of $\text{Stab}(x)$ is the number of left cosets of $\text{Stab}(x)$, which by Theorem 1.33 is the size of the orbit of $x$. □

**Example 1.35.** Consider the graph $\Gamma$ shown in Figure 1.7. It is not difficult to argue that the action of $\text{SYM}(\Gamma)$ is vertex transitive, hence for any vertex $v$, $\text{Orb}(v)$ consists of all six vertices. It is also not too difficult to argue that $\text{Stab}(v) \approx D_4$ for any vertex $v$. Thus $|\text{SYM}(\Gamma)| = 8 \cdot 6 = 48$.

Another insight provided by Theorem 1.33 is the following method of identifying elements in a group with the elements in the orbit of an $x$ where $\text{Stab}(x)$ is trivial.

**Corollary 1.36.** *Let $G \curvearrowright X$. For any $x \in X$, where $\text{Stab}(x) = e$, there is a one-to-one correspondence between the elements of $G$ and the elements of $\text{Orb}(x)$.*

## 1.5 Generating Sets and Cayley Graphs

We are now in a position to present a useful extension of Cayley's Basic Theorem that applies to finitely generated groups.

### 1.5.1 Generators

**Definition 1.37.** If $G$ is a group and $S$ is a subset of elements, then $S$ *generates* $G$ if every element of $G$ can be expressed as a product of elements from $S$ and inverses of elements of $S$. A group $G$ is *finitely generated* if it has a finite generating set.

**Example 1.38.** Every finite group is finitely generated; take $S = G$ and you satisfy the definition. However, it is sometimes useful to know a natural, proper subset of the elements that generate the whole group. For example, it is often demonstrated in an introductory group theory course that the adjacent transpositions $S = \{(12), (23), \ldots, (n-1\ n)\}$ generate $\text{SYM}_n$. Another well-known generating set consists of a transposition and an $n$-cycle: $S = \{(12), (12 \cdots n)\}$.

**Example 1.39.** One may be tempted to think of generating sets as analogous to bases of a vector space. This intuition is not horrible, but it is not perfect either. For example, the group of integers is generated by the single element $1 \in \mathbb{Z}$. One can form a larger and redundant set such as $S = \{1, 2\}$, which also forms a generating set. Of course this set can be "trimmed" to form a minimal generating set, just as a set of vectors which spans a vector space can be "trimmed" to a basis. However, there are other sets such as $\{2,3\}$ that generate $\mathbb{Z}$ and that can not be made smaller by simply removing a single element. Similarly one can show that $\{6,10,15\}$ is a generating set for $\mathbb{Z}$, but no proper subset generates $\mathbb{Z}$.

**Example 1.40.** Any element of the dihedral group of order $2n$, $D_n$, is determined by where it takes two adjacent vertices of the associated regular $n$-gon. We use this fact to establish that $D_n$ is generated by a rotation and a reflection.

Let • and ◦ denote adjacent vertices of the $n$-gon (as in the hexagon of Figure 1.10) and let $g$ be any element of $D_n$. The element $g$ takes • to $g(\bullet)$ and ◦ to $g(\circ)$. Let $\rho$ be a rotation – clockwise or counterclockwise, it is your choice – through an angle of $2\pi/n$. Let $\phi$ be the reflection that fixes •. Then there is a number $m$ such that $\rho^m(\bullet) = g(\bullet)$. If it happens that $\rho^m(\circ) = g(\circ)$, we are done, since in this case, $g = \rho^m$. Said another way, since $g^{-1}\rho^m$ fixes • and ◦, $g^{-1}\rho^m = e \Rightarrow \rho^m = g$. Otherwise the element $\rho^m \cdot \phi$ takes • $\rightarrow g(\bullet)$ and ◦ $\rightarrow g(\circ)$. Thus $g^{-1}\rho^m\phi = e$ and therefore $\rho^m\phi = g$. Since $g$ was arbitrary, we have shown that any element of $D_n$ can be expressed as a product of $\rho$'s and $\phi$'s.

One can also generate $D_n$ using two adjacent reflections. Let $\phi$ be as above and let $\psi$ be a reflection of the $n$-gon that forms an angle of $\pi/n$ with $\phi$. Then $\phi \cdot \psi$ is a rotation through an angle of $2\pi/n$. By the previous argument, $\phi$ and the product $\phi \cdot \psi$ generate $D_n$, hence $\phi$ and $\psi$ generate $D_n$.

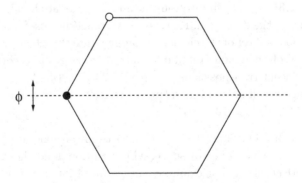

Fig. 1.10. A regular hexagon with two distinguished adjacent vertices.

As you can see in the examples above, a given group can have different, interesting, and fairly distinct generating sets. Although it is a guideline that is frequently broken, it is best to not refer to a particular set of generators as *the* generators of a group. Authors often circumvent this issue using phrases such as "the natural generating set" or "the generating set we are considering."

We end this list of examples with a group where it is easy to show that the group admits no finite set of generators.

**Example 1.41.** The group of rational numbers under addition is not finitely generated. To see this, assume to the contrary that $\mathbb{Q}$ is generated by $S = \left\{ \dfrac{n_1}{d_1}, \dfrac{n_2}{d_2}, \dots, \dfrac{n_k}{d_k} \right\}$. If $d$ is the least common multiple of the denominators, that is $d = \mathrm{LCM}[d_1, d_2, \dots, d_k]$, then every element that can be written as a sum of the supposed generators and their inverses can be expressed as $\dfrac{n}{d}$ for some $n \in \mathbb{Z}$. But not every element in $\mathbb{Q}$ has such an expression.

### *1.5.2 Cayley's Better Theorem*

The improvement of Cayley's Basic Theorem is:

**Theorem 1.42** (Cayley's Better Theorem). *Every finitely generated group can be faithfully represented as a symmetry group of a connected, directed, locally finite graph.*

Just as in Cayley's Basic Theorem it is hard, initially, to see how one might prove Cayley's Better Theorem. One has a group $G$ and some finite generating set $S$, and somehow out of whole cloth one must produce both a graph and an action of $G$ on that graph.

*Proof of Cayley's Better Theorem.* We construct a graph $\Gamma_{G,S}$ using the group $G$ and the generating set $S$. The vertices of $\Gamma_{G,S}$ are the elements of $G$. For each $g \in G$ and each $s \in S$ form a directed edge with initial vertex $g$ and terminal vertex $g \cdot s$. Thus edges of $\Gamma_{G,S}$ correspond to right multiplication by elements of the generating set $S$.

To show that the graph is connected it suffices to show that one can get from the vertex corresponding to the identity $e$ to any other vertex $g$. (Explain to yourself why this is sufficient!) Since $S$ is a generating set, we can write $g$ as a product of generators and their inverses $g = s_1 s_2 \cdots s_n$, where each $s_i$ is in $S$ or $S^{-1}$. In the graph $\Gamma_{G,S}$ there is an associated path:

$$\bullet \xrightarrow{s_1} \bullet \xrightarrow{s_2} \bullet \xrightarrow{s_3} \cdots \xrightarrow{s_n} \bullet$$
$$e \qquad s_1 \qquad s_1 s_2 \qquad\qquad s_1 s_2 \cdots s_n = g$$

Here we have been sloppy, and have not bothered to take orientation into account. If any of the letters $s_i$ in the expression $g = s_1 s_2 \cdots s_n$ is the inverse of an element of $S$, then one would be travelling opposite to the orientation of the directed edge joining $s_1 s_2 \cdots s_{i-1}$ to $s_1 s_2 \cdots s_{i-1} s_i$.

To avoid confusion between elements of $G$ and vertices in $\Gamma_{G,S}$, let the vertex in $\Gamma_{G,S}$ associated to $g$ be denoted $v_g$.

The graph $\Gamma_{G,S}$ is locally finite because the set $S$ is finite. In fact, each and every vertex will be incident with exactly $2|S|$ edges. Then $v_g$ is the initial vertex of exactly one directed edge labelled $s$ for each $s \in S$ (going from $v_g$ to $v_{g \cdot s}$) and it is the terminal vertex of exactly one directed edge labelled $s$ for each $s \in S$ (going from $v_{g \cdot s^{-1}}$ to $v_g$).

The proof of Cayley's Basic Theorem describes the left action of $G$ on the vertices of this graph: the element $g \in G$ sends the vertex $v_h$ to $v_{g \cdot h}$.

Does this action on vertices extend to an action on the edges of $\Gamma_{G,S}$? Notice that the vertex $v_h$ is joined to $v_{h \cdot s}$ by a directed edge labelled $s$. The element $g \in G$ takes the vertex $v_h$ to $v_{g \cdot h}$ and the vertex $v_{h \cdot s}$ to $v_{g \cdot h \cdot s}$. Thus we may define the action of $g \in G$ on the edges by stating the edge labelled $s$ joining $v_h$ to $v_{hs}$ is sent to the edge labelled $s$ joining $v_{gh}$ to $v_{ghs}$. (See Figure 1.11.) $\qquad\qquad \square$

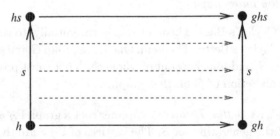

Fig. 1.11. The action of an element $g \in G$ on the vertices of $\Gamma_{G,S}$ extends to an action on the edges.

We note that it is precisely because the group action is on the left, while the edges are defined in terms of right multiplication, that the action of $G$ on the vertices can be extended to the edges of $\Gamma_{G,S}$. In other words: if $s$ is a generator, then there is an edge labelled $s$ joining $v_h$ to $v_{hs}$; if we left multiply by $g \in G$ then $v_h$ and $v_{hs}$ go to $v_{gh}$ and $v_{ghs}$, respectively; but then by definition there is an edge labelled $s$ joining $v_{gh}$ to $v_{ghs}$. This is illustrated in Figure 1.11.

**Definition 1.43.** Given a group $G$ and a finite generating set $S$, call the directed graph $\Gamma_{G,S}$ the *Cayley graph of $G$ with respect to $S$*, or just the *Cayley graph* when the group $G$ and generating set $S$ are sufficiently clear from context.

We note that the construction of the action of $G$ on the Cayley graph $\Gamma_{G,S}$ produces an action that preserves the orientation of the directed edges and the labellings on those edges. Thus $G < \text{Sym}^+(\Gamma_{G,S})$.

**Example 1.44.** In Figure 1.12 we show the Cayley graph of $\text{Sym}_3$ with respect to the generating set $\{(12), (123)\}$. The version on the left follows the definition precisely. However, we notice that when a generator is an involution (an element of order 2) there is little need to have a "parallel" pair of directed edges. It is a standard convention to replace the two directed edges with a single undirected edge, as is done in the figure on the right. While this is not what is described by the definition of a Cayley graph, it is often more aesthetically pleasing and easier to understand. When this convention is employed, the claim that every vertex is of degree $2|S|$ is no longer true. Further, the graph is then a strange hybrid of directed and undirected edges.

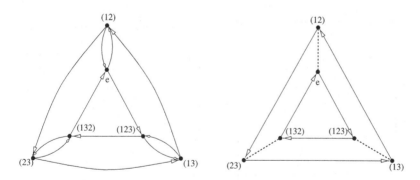

Fig. 1.12. The Cayley graph of $\text{Sym}_3$ with respect to $\{(12), (123)\}$, drawn using two different conventions on how to treat generators of order 2.

**Exercise 1.45.** You may have noticed that Figures 1.1 and 1.12 are quite similar. However, the vertices corresponding to group elements have been slightly shifted. Figure out what is similar and what is different in these two figures. Along the way, determine what the action of SYM₃ on its Cayley graph looks like.

**Remark 1.46.** Arthur Cayley introduced Cayley graphs in a paper published in 1878 [Ca78]. In that paper he mentions that when there are two or more generators "we require different colours, the [edges] belonging to any one substitution [i.e. generator] being of the same colour." It is often quite helpful, if you own enough colored pens or pencils, to draw Cayley graphs using this sort of color convention.

## 1.6 More Cayley Graphs

The best way to understand a new definition is to look at a few nontrivial examples.

### *1.6.1 Dihedral Groups*

Let $D_n$ be the dihedral group of order $2n$. As was discussed in Example 1.40, there are two standard generating sets for $D_n$: the set consisting of a reflection $\phi$ and a minimal rotation $\rho$ or a set consisting of two reflections $\{\phi, \psi\}$ where the angle between the fixed lines of these reflections forms an angle of $\pi/n$. The corresponding Cayley graphs, in the case where $n = 4$, are shown in Figure 1.13.

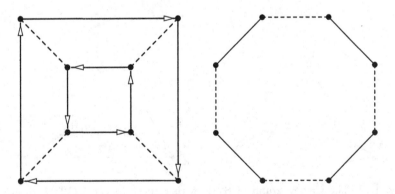

Fig. 1.13. Two Cayley graphs of $D_4$. On the left is the graph with respect to a generating set consisting of a rotation and a reflection; on the right the generating set consists of two adjacent reflections.

One can construct these Cayley graphs directly from the action of $D_n$ on a regular $n$-gon. In Figure 1.14, on the left, we have chosen a point whose stabilizer is trivial, so that the elements in the orbit of this point correspond to the six elements of $D_3$ (Corollary 1.36). We have also indicated two particular reflections to be used as the generating set of $D_3$. One can then connect the points in the orbit to form the Cayley graph with respect to the indicated generators. We start with our original point, the only non-solid point in the figure, and apply the two generators to it, in order to get the intermediate figure. If $\circ$ denotes our original point, then the top-right point is $a \cdot \circ$. By definition, in the Cayley graph this point is joined by a dashed edge to the vertex $ab \cdot \circ$. Noticing that $ab$ is a clockwise rotation through an angle of $120°$, one can begin to verify that the picture shown on the right in Figure 1.14 is indeed accurate.

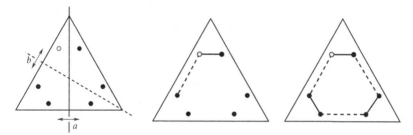

Fig. 1.14. The dihedral group $D_3$ is generated by the reflections $a$ and $b$. The associated Cayley graph can be constructed by connecting points in an orbit of a point in "general position."

This particular example, where $n = 3$, is easily generalized to the case where one is using two adjacent reflections to generate $D_n$.

**Exercise 1.47.** Follow a procedure similar to the one outlined above to find an embedding of the Cayley graph of $D_n$, with respect to a reflection and a rotation through an angle of $2\pi/n$, inside a regular $n$-gon. You will want to make a judicious choice for the initial point whose orbit corresponds with the elements of $D_n$.

### 1.6.2 Symmetric Groups

At the end of the previous section we produced the Cayley graph of SYM$_3$ with respect to the generating set $\{(12), (123)\}$. Here we will examine the Cayley graph of SYM$_4$ with respect to the generating set $\{(12), (23), (34)\}$. This Cayley graph has $4! = 24$ vertices, where each vertex is incident to three edges, using the convention that involutions are represented by unoriented edges. The products $(12)(23) = (123)$ and $(23)(34) = (234)$ are both of order 3, while the transpositions $(12)$ and $(34)$ commute. It follows that at each vertex of $\Gamma$ the edges corresponding to $(12)$ and $(23)$ will determine a 6-cycle, the edges corresponding to $(23)$ and $(34)$ give another 6-cycle, while the edges corresponding to $(12)$ and $(34)$ determine a 4-cycle. Thus at each vertex of this Cayley graph one sees two hexagons and a square. This is enough data to determine that this Cayley graph has the form indicated in Figure 1.15.

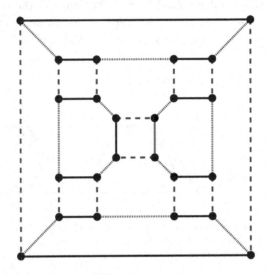

Fig. 1.15. The Cayley graph of the symmetric group SYM$_4$ with respect to the generating set $S = \{(12),(23),(34)\}$.

If you have studied the Archimedean solids you might notice that this Cayley graph looks like the edges of a truncated octahedron. This is no accident. The following argument – similar to the one that embedded Cayley graphs of dihedral groups into the interior of regular $n$-gons – explains why.

First, represent SYM$_4$ as the group of symmetries of the regular tetrahedron $\mathbb{T}$. To do this, identify the four vertices of $\mathbb{T}$ with the numbers $[4] = \{1,2,3,4\}$. Any symmetry of $\mathbb{T}$ results in a permutation of $[4]$, which is how one constructs a homomorphism $\phi : \text{SYM}(\mathbb{T}) \to \text{SYM}_4$.

One can prove that $\phi$ is onto by showing that the three transpositions $(12),(23)$ and $(34)$ are in the image. The reflection across the plane spanned by the edge joining vertex 3 to vertex 4, and the midpoint of the edge joining 1 to 2, is taken by $\phi$ to the transposition $(12)$. (See

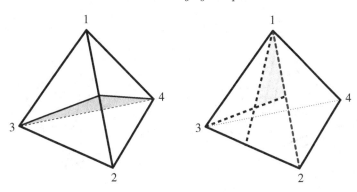

Fig. 1.16. The group of symmetries of a regular tetrahedron is isomorphic to SYM₄. On the left, the shaded portion indicates the fixed points for the reflection of $\mathbb{T}$ corresponding to the transposition (12). On the right, portions of the fixed point set for reflections corresponding to (12), (23) and (34) are indicated by dashed lines.

Figure 1.16.) Similar reflections yield the generating set for SYM₄. Because the two groups both have order 24, and $\phi$ is onto, it must be the case that $\phi$ is an isomorphism.

To draw the Cayley graph with respect to $S = \{(12), (23), (34)\}$ we pick a point in $\mathbb{T}$ whose stabilizer is trivial. By drawing in some of the fixed sets for the reflections corresponding to these transpositions, we see a triangular region on the exterior of the tetrahedron, shown on the right in Figure 1.16, any of whose points fit the bill. Picking a point, and applying the reflections corresponding to (12) and (23) we see a hexagonal portion of this Cayley graph. (This subgraph illustrates the fact that this subgroup is isomorphic to $D_3$.) Using the same point and the reflections corresponding to (12) and (34) we see a rectangular portion of the Cayley graph. These are highlighted in Figure 1.17. The hexagon corresponding to the generators (23) and (34) is not shown in this figure, but it sits just a bit below the vertex labelled 1. From these facts it is not hard to establish that the Cayley graph we are constructing has its underlying structure coming from the edges of an Archimedean solid.

**Exercise 1.48** (hard!). Give a description of the Cayley graph of SYM$_n$, with respect to the generating set $S = \{(12), (23), \ldots (n-1\ n)\}$, for all $n \geq 5$.

**Remark 1.49** (Drawing Trick). In the previous examples we used a common technique for discovering the structure of – also known as

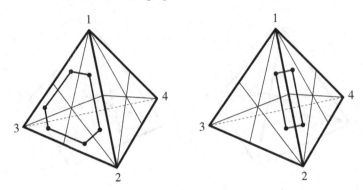

Fig. 1.17. Parts of the Cayley graph of SYM4 shown embedded in the tetrahedron.

"figuring out how to draw" – a Cayley graph. We started with an action
of the group we were interested in on a geometric object ($D_n$ acting on
a regular $n$-gon and SYM4 acting on the tetrahedron), then we found a
point in that space that is moved freely by the group action. By Corollary 1.36, there is a bijection between the orbit of this point under the
action of $G$ and the elements of $G$. Thus we may use the orbit of the point
as the vertex set of the Cayley graph. Our point was sufficiently well-
chosen that the edges ended up embedding nicely in the space. While
this method is not guaranteed to be useful in all situations, it is quite
often helpful for visualizing Cayley graphs.

### 1.6.3 The Symmetry Group of a Cube

Let $\square$ denote a cube, and let $G = \mathrm{Sym}(\square)$ be the group of all symmetries
of $\square$. In order to be very concrete, let $\square$ be thought of as the convex hull
of the eight points in $\mathbb{R}^3$ whose coordinates are of the form $(\pm 1, \pm 1, \pm 1)$,
and let SYM($\square$) consist of all reflections and rotations that take $\square$ back
to itself.

The order of $G$ can be computed using the Orbit-Stabilizer Theorem
(Corollary 1.34). Let $v$ be the vertex at $(1, 1, 1)$. Then the orbit of $v$
consists of the eight vertices of $\square$. Since any element in the stabilizer of
$v$ permutes the three edges incident with $v$, and no non-trivial symmetry
fixes these three edges, we have $\mathrm{Stab}(v) \approx \mathrm{SYM}_3$. Thus

$$|G| = |\mathrm{Orb}(v)| \cdot |\mathrm{Stab}(v)| = 8 \cdot 6 = 48.$$

The insight behind the Orbit-Stabilizer Theorem also gives us a method for finding a nice set of generators for $G$. Reflections in the coordinate planes are all that are needed to move $v$ to any other vertex of $\square$. Reflections in the appropriate planes of the form $x = \pm y$, $x = \pm z$, and $y = \pm z$ generate $\mathrm{Stab}(v)$ for any vertex $v$. We can put all of these reflecting planes into a picture of the cube, as in Figure 1.18. This shows the surface of the cube has been dissected into 48 triangles. Fix any one of these triangles. It can be checked by hand that the three reflections in the sides of the chosen triangle generate $G$. (This is an idea we discuss a bit more formally in Section 1.8.)

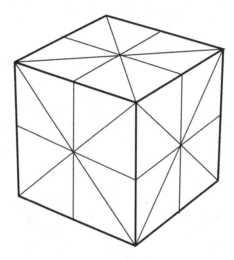

Fig. 1.18. A cube with the fixed planes for reflections indicated.

To find the Cayley graph, we use the Drawing Trick, using any point inside our chosen triangle. The reader should check the following claims: The portion of the Cayley graph around each vertex of $\square$ forms a hexagon; each edge of $\square$ has a corresponding square in the Cayley graph; and each face of $\square$ has an octagon corresponding to the $D_4$ stabilizer of that face. From this one sees that the Cayley graph is the edge set of the great rhombicuboctahedron (see Figure 1.19).

### 1.6.4 Free Abelian Groups

Introductions to group theory rarely feature extended field trips to the Zoo of Infinite Groups. The few times that they do venture into the Zoo,

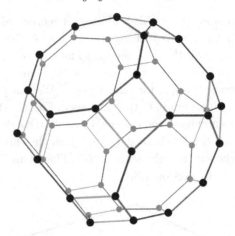

Fig. 1.19. A Cayley graph of the symmetry group of a cube.

they stay well away from the more exotic exhibits. We will encounter some of these later in the text but, for now, let's just look at one of the basic beasts.

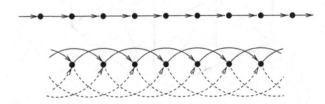

Fig. 1.20. Two Cayley graphs of $\mathbb{Z}$. The top is with respect to $S = \{1\}$, the bottom with respect to $S' = \{2,3\}$.

The Cayley graph of $\mathbb{Z}$ with respect to the single cyclic generator $\{1\}$ is a combinatorial version of the real line. If one uses a generating set such as $\{2,3\}$ the Cayley graph is a bit more complicated, but somehow it still seems to keep the basic "long and skinny" shape of the real line. (See Figure 1.20.)

If you generate the free abelian group $\mathbb{Z}^n$ using "the standard basis vectors"

$$S = \{(1,0,0,\ldots,0),(0,1,0,\ldots,0),\ldots,(0,0,0,\ldots,1)\}$$

the resulting Cayley graph consists of the edges of a cubic lattice in $\mathbb{R}^n$. The vertices correspond to points whose coordinates are all integers; the

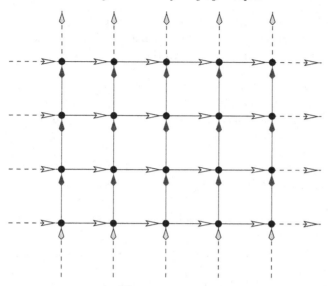

Fig. 1.21. The Cayley graph of $\mathbb{Z} \oplus \mathbb{Z}$ with respect to the generators $\{(1,0),(0,1)\}$.

edges connect points that are exactly a distance 1 apart. The case where $n = 2$ is shown in Figure 1.21.

**Remark 1.50.** Cayley restricted his attention to the Cayley graphs of finite groups, and this restriction also occurs in many elementary presentations of Cayley graphs. In 1912 Max Dehn demonstrated the utility of Cayley graphs, and a related construction called *Gruppenbilder*, in the study of finitely generated infinite groups [De12]. This work of Max Dehn is often cited as the start of the combinatorial and geometric study of finitely generated infinite groups.

## 1.7 Symmetries of Cayley Graphs

Having defined symmetry groups of graphs, one might wonder if there is any nice description of the groups that arise as the symmetry groups of graphs. The following theorem, first proved by Frucht in 1938, indicates that all finitely generated groups can be realized as label and orientation preserving symmetries of locally finite, directed graphs. In Exercise 20, at the end of this chapter, you are asked to show even more: *Every finitely generated group is realizable as the full symmetry group of a locally finite graph.* Thus knowing that a group $G$ acts on a graph $\Gamma$

does not immediately imply anything about the group $G$. On the other hand, this result also shows that viewing groups via their actions on graphs is a very flexible approach to the study of group theory.

**Theorem 1.51.** *Let $\Gamma_{G,S}$ be the Cayley graph of a group $G$ with respect to a finite generating set $S$. Consider $\Gamma_{G,S}$ to be decorated with directions on its edges and labellings of its edges, corresponding to the generating set $S$. Then $\textsc{Sym}^+(\Gamma_{G,S}) \approx G$.*

*Proof.* The left action of $G$ on $\Gamma_{G,S}$ given by Cayley's Better Theorem (1.42) shows that $G \hookrightarrow \textsc{Sym}(\Gamma_{G,S})$. Since this is a left action, it does not affect the directions or the labellings on the edges of $\Gamma_{G,S}$, which are defined in terms of right multiplication. Thus $G \hookrightarrow \textsc{Sym}^+(\Gamma_{G,S})$. To show that this map is surjective, we consider an arbitrary element $\gamma \in \textsc{Sym}^+(\Gamma_{G,S})$. For any element $g \in G$ let $\bullet_g$ be the vertex in $\Gamma_{G,S}$ corresponding to $g$. Then there is a $g$ such that $\gamma(\bullet_e) = \bullet_g$. Considering $g$ as a symmetry of $\Gamma_{G,S}$, the product $\gamma \cdot g^{-1} \in \textsc{Sym}^+(\Gamma_{G,S})$. This symmetry takes $\bullet_e$ to $\bullet_e$. Further, it fixes all the edges arriving at or leaving from $e$, as elements of $\textsc{Sym}^+(\Gamma_{G,S})$ preserve direction and labelling. Hence $\gamma \cdot g^{-1}$ fixes all of the vertices adjacent to $\bullet_e$. But then, again, it will fix edges incident to these vertices and so on. So the symmetry $\gamma \cdot g^{-1}$ is the identity, which is to say, $\gamma = g$ (as elements in $\textsc{Sym}^+(\Gamma_{G,S})$). Since $\gamma$ was an arbitrary symmetry of the Cayley graph, our homomorphism is a bijection, and $G \approx \textsc{Sym}^+(\Gamma_{G,S})$. $\qquad\square$

## 1.8 Fundamental Domains and Generating Sets

Let $G$ be a group acting on a connected graph $\Gamma$. In this section we show how one can form a fundamental domain for the action, and then given a fundamental domain, how one can find a generating set for $G$. This technique is one example of how an understanding of a group action on some geometric object can result in information about the group itself.

In the lemma below we have used the adjective "closed." This term does have a precise mathematical meaning, which you may have encountered in a course on real analysis or topology. If you have not previously encountered this word, you may take "closed" to mean "containing its end points" as in the interval $[0,1]$ is a closed subset of $\mathbb{R}$, but $[0,1)$ and $(0,1)$ are not. Having a deep understanding of closed sets is not necessary to follow the arguments given here.

**Lemma 1.52.** *If a group $G$ acts on a connected graph $\Gamma$ then there is a subset $\mathcal{F} \subset \Gamma$ such that:*

1. *$\mathcal{F}$ is closed;*
2. *the set $\{g \cdot \mathcal{F} \mid g \in G\}$ covers the graph $\Gamma$; and*
3. *no subset of $\mathcal{F}$ satisfies properties (1) and (2).*

The subset $\mathcal{F}$ described in Lemma 1.52 is called a *fundamental domain* for the action $G \curvearrowright \Gamma$. Two examples of fundamental domains are shown in Figure 1.22. We establish Lemma 1.52 by building up a subset of $\Gamma$, starting with a single vertex, and then adding to it until it is sufficiently large so that its image under the action of $G$ covers the entire graph $\Gamma$. The actual proof of Lemma 1.52 is long, and is probably the most challenging argument presented in this first chapter. The reader is strongly encouraged to use all of his or her favorite techniques for understanding complicated arguments as they proceed through the details.

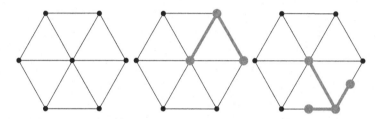

Fig. 1.22. A wheel graph $\Gamma$ and two examples of fundamental domains for the action of $\mathbb{Z}_6$ on $\Gamma$, the action being by rotations. Any fundamental domain for this action must include the central vertex but, after that, there are many choices as to how to construct a fundamental domain.

*Proof of Lemma 1.52.* To construct a fundamental domain, start with a fixed vertex $v \in \Gamma$, and consider connected subgraphs $\mathcal{C} \subset \Gamma$ where

1. $v \in \mathcal{C}$; and
2. if $x$ and $y$ are distinct vertices in $\mathcal{C}$ then there is no element $g \in G$ such that $g \cdot x = y$.

The vertex $v$ by itself satisfies these two conditions, so there are such subgraphs in $\Gamma$. Let $\mathcal{C}_0 \subset \mathcal{C}_1 \subset \mathcal{C}_2 \subset \cdots$ be a sequence of subgraphs satisfying our conditions, one properly contained in the next. If there are only finitely many orbits of vertices under the action of $G$, and the graph $\Gamma$ is locally finite, then at some point one of the $\mathcal{C}_i$'s in the sequence above will be a maximal subgraph satisfying the conditions above. In this case, call this subgraph the CORE. Otherwise form a maximal subgraph satisfying our conditions by defining

$$\text{CORE} = \bigcup \mathcal{C}_i$$

where the sequence of $\mathcal{C}_i$'s is as large as it can possibly be.[4] It is perhaps less clear that CORE satisfies our conditions (1) and (2) in this case. But notice that:

1. $v \in \mathcal{C}_0 \subset$ CORE; and
2. if $x$ and $y$ are distinct vertices in CORE then there must be some $\mathcal{C}_i$ that contains them. Thus there is no $g \in G$ taking $x$ to $y$.

The subgraph CORE does not necessarily form a fundamental domain for the action of $G$ on $\Gamma$, but it does contain all the vertices needed.

Claim: *The image of* CORE *under the action of* $G$ *contains all the vertices of* $\Gamma$.

Assume to the contrary that there is a vertex $v \in \Gamma$ that is not contained in some $g \cdot$ CORE, $g \in G$. Let $\{v = v_0, v_1, \dots, v_n\}$ be the vertices in a minimal-length edge path joining $v$ to a vertex in $G \cdot$ CORE. It must be the case that $v_{n-1}$ is not in $G \cdot$ CORE (since otherwise there would be a shorter path to $G \cdot$ CORE), which implies that $v_{n-1}$ is a vertex outside of $G \cdot$ CORE which is joined to $G \cdot$ CORE by a single edge. So we may assume that $v \in \Gamma \setminus G \cdot$ CORE and there is an edge $e$ joining $v$ to some vertex in $g \cdot$ CORE for some $g \in G$. (In other words, we can now assume "$v_{n-1}$" is "$v$.") But then we could add $g^{-1} \cdot v$ and $g^{-1} \cdot e$ to CORE to create a larger subgraph satisfying the properties listed above, contradicting the assumption that CORE is a maximal subgraph with these properties. Thus if $v \in \Gamma$ then $v \in g \cdot$ CORE for some $g \in G$.

To complete the proof of Lemma 1.52, we need to (possibly) expand the CORE, so that the image of CORE under the action of $G$ covers every edge. An example where the CORE cannot be a fundamental domain by itself comes from Cayley graphs. The action of a group $G$ on its Cayley graph $\Gamma$ is vertex transitive. From this it follows that CORE consists of the single vertex $v$, and therefore $G \cdot$ CORE is the vertex set of $\Gamma$, hence $G \cdot$ CORE contains no edges. If there are edges in $\Gamma$ that are not in $G \cdot$ CORE, we form a fundamental domain $\mathcal{F}$ by adding half-edges to CORE. Assume then that we have an edge $e \not\subset G \cdot$ CORE where

---

[4] Here we are running into some dicey logical/set-theoretic issues, directly connected to the axiom of choice.

ENDS($e$) ∩ CORE ≠ ∅. If ENDS($e$) ⊂ CORE, then we could have added $e$ to CORE and still satisfied our conditions above. So we may assume that ENDS($e$) = $\{v, w\}$ with $v \in$ CORE and $w \notin$ CORE. Let $h_e$ be the closed half of $e$ that contains $v \in$ CORE. Define $\mathcal{F}$ to be the union of CORE and these half-edges. In the case of a group acting on its Cayley graph, $\mathcal{F}$ is a "star of half-edges," consisting of a single vertex $v$ and every half-edge $h_e$ where ENDS($e$) contains $v$.

Since $\mathcal{F} \supset$ CORE, V($\Gamma$) ⊂ $G \cdot \mathcal{F}$. It remains only to show that the edges of $\Gamma$ are also in $G \cdot \mathcal{F}$. To that end, let $e$ be an edge in $\Gamma$ where ENDS($e$) = $\{v, w\}$. If $e \notin G \cdot$ CORE then there is a $g$ and $\hat{g} \in G$ such that $v \in g \cdot$ CORE and $w \in \hat{g} \cdot$ CORE. Thus $e$ will be covered by the $g$ and $\hat{g}$ images of the half-edges added to CORE in creating $\mathcal{F}$. □

**Example 1.53.** Let $\Gamma = K_{n,m}$ with $n \neq m$. Let $G =$ SYM($\Gamma$) and note that $G$ acts transitively on $V_\circ$ and on $V_\bullet$. Thus CORE can be taken to be any edge of $\Gamma$, and one does not need to add half-edges to CORE in order to form a fundamental domain.

On the other hand, if $n = m$ then there is a symmetry that exchanges $V_\circ$ and $V_\bullet$. In this case CORE consists of a single vertex, and a fundamental domain is formed by adding a half-edge that contains this vertex.

**Example 1.54.** The symmetry group of the graph $\Gamma$ shown in Figure 1.23 is isomorphic to $D_3$. This group acts transitively on the set of 4-valent vertices and on the set of 2-valent vertices, but no symmetry of $\Gamma$ takes a vertex of valence 2 to a vertex of valence 4. Thus CORE can be taken to be any edge joining a vertex of valence 2 to a vertex of valence 4. However, the image of such an edge under the action of SYM($\Gamma$) is just the hexagon on the outside of the graph; it does not include the interior edges. Thus a half-edge needs to be added to CORE in order to form a fundamental domain. This is indicated in the picture on the right of Figure 1.23.

**Theorem 1.55.** *Let $G$ act on a connected graph $\Gamma$ with fundamental domain $\mathcal{F}$. Then the set of elements*

$$S = \{g \in G \mid g \neq e \text{ and } g \cdot \mathcal{F} \cap \mathcal{F} \neq \emptyset\}$$

*is a generating set for $G$.*

*Proof.* Let $g \in G$ be an arbitrary element and let $v$ be a vertex in the fundamental domain $\mathcal{F}$. Choose a path $p$ joining $v$ to $g \cdot v$ and

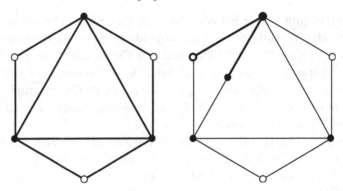

Fig. 1.23. A bipartite graph $\Gamma$ where there are two orbits of vertices under the action of $\text{Sym}(\Gamma)$. A fundamental domain for the action $\text{Sym}(\Gamma) \curvearrowright \Gamma$ is shown on the right.

let $\{\mathcal{F}, g_1\mathcal{F}, g_2\mathcal{F}, \ldots, g_n\mathcal{F} = g\mathcal{F}\}$ be a finite sequence of images of the fundamental domain such that

1. the entire path $p$ is contained in $\bigcup g_i\mathcal{F}$, and
2. $g_i\mathcal{F} \cap g_{i+1}\mathcal{F} \neq \emptyset$, where $g_0$ is defined to be the identity.

Since $\mathcal{F} \cap g_1\mathcal{F} \neq \emptyset$, $g_1 \in S$ by definition. Similarly $g_1\mathcal{F} \cap g_2\mathcal{F} \neq \emptyset$ implies $\mathcal{F} \cap g_1^{-1}g_2\mathcal{F} \neq \emptyset$, thus $g_1^{-1}g_2 \in S$ and so $g_2$ is a product of elements in $S$. Continuing in this manner we see that $g_n = g$ must also be a product of elements of $S$. $\qquad\square$

**Exercise 1.56.** Let $\Gamma$ be the graph shown in Figure 1.23 and let $\mathcal{F}$ be the indicated fundamental domain for the action of $\text{Sym}(\Gamma) \approx D_3$. Show that the generating set given by Theorem 1.55 consists of two adjacent reflections.

**Example 1.57.** If one applies Theorem 1.55 to the action $\text{Sym}_5 \curvearrowright K_5$, the resulting set of generators is quite large. Label the vertices of a complete graph $K_5$ by the numbers 1 through 5 (to make the isomorphism $\text{Sym}_5 \approx \text{Sym}(K_5)$ explicit) and take the fundamental domain to be the one indicated in Figure 1.24. Let $\text{Sym}_{\{1,2,3,4\}}$ be the copy of the symmetric group $\text{Sym}_4$ that acts on $K_5$, fixing the vertex labelled 5. The image of the chosen fundamental domain under the action of any element of $\text{Sym}_{\{1,2,3,4\}}$ intersects the fundamental domain at least at the vertex labelled 5, so this entire subgroup (except the identity element) is contained in $S$. Continuing to abuse notation we note that

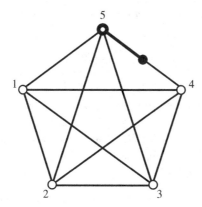

Fig. 1.24. A fundamental domain for the action of SYM$_5$ on the complete graph with five vertices.

the only other elements in $S$ can be expressed as a transposition $(45)$ combined with some (possibly trivial) permutation from SYM$_3$. There are six such symmetries, so $|S| = 23 + 6 = 29$.

The argument above applies to the action of SYM$_n$ on $K_n$, giving a huge generating set. However, an induction argument based on these actions shows that the subset $\{(12), (23), \ldots, (n-1, n)\}$ is also a generating set for SYM$_n$. The base case consists of checking that $(12)$ generates SYM$_2$; the induction step assumes $\{(12), (23), \ldots, (n-1, n)\}$ generates SYM$_n$, and then notes that the generating set given by Theorem 1.55 for the action $\Sigma_{n+1} \curvearrowright K_{n+1}$ consists of the non-identity elements in SYM$_n$ and the product of the transposition $(n, n+1)$ with elements of SYM$_{n-1}$. Thus adding $(n, n+1)$ to the generating set for SYM$_n$ is all that is needed to generate SYM$_{n+1}$.

In addition to finding generating sets, fundamental domains can be useful in computing the index of subgroups.

**Theorem 1.58.** *Let $G \curvearrowright \Gamma$ with fundamental domain $\mathcal{F}$. Further assume that if $g \cdot \mathcal{F} = \mathcal{F}$ then $g = e \in G$. If $H < G$ and a fundamental domain for the induced action $H \curvearrowright \Gamma$ is a union of $n$ copies of $\mathcal{F}$ (where $n \in \mathbb{N} \cup \{\infty\}$) then the index of $H$ in $G$ is $n$.*

*Proof.* This is essentially an application of Theorem 1.33. Since no non-identity element of $G$ fixes $\mathcal{F}$, there is a point $x$ in $\mathcal{F}$ that is moved freely under the action of $G$. Thus there is a bijection between the elements of $G$ and points in $\mathrm{Orb}(x)$. Express the fundamental domain for $H \curvearrowright \Gamma$ as an explicit union of copies of $\mathcal{F}$:

$$\mathcal{F}_H = \bigcup_{i=1}^{n} g_i \mathcal{F}$$

where the $g_i$ are distinct elements of $G$ and $n \in \mathbb{N} \cup \{\infty\}$.

Since $\mathcal{F}_H$ is a fundamental domain for the action of $H$, if $g \in G$ then $g \cdot x = g_i h \cdot x$ for one of the $g_i$ above. Thus $G = \bigcup_{i=1}^{n} g_i H$. Further, if $g_i H \cap g_j H \neq \emptyset$, then $g_i g_j^{-1} \in H$, which contradicts the fact that $\mathcal{F}_H$ is a fundamental domain. Thus the $g_i$ form a set of coset representatives for $H$ in $G$ and $[G : H] = n$. $\qquad\square$

**Example 1.59.** Let $C_8$ be an 8-cycle, that is, a connected graph with eight vertices joined in a single cycle. The symmetry group of $C_8$ is $D_8$, and a fundamental domain for $D_8 \curvearrowright C_8$ consists of a single vertex and half-edge. There is a cyclic subgroup $\mathbb{Z}_4 < \mathrm{SYM}(C_8)$, which can be visualized as rotations through an angle of $90n°$. A fundamental domain for $\mathbb{Z}_4 \curvearrowright C_8$ consists of a combinatorial arc consisting of three vertices joined by two edges. (See Figure 1.8.) As the fundamental domain for $\mathbb{Z}_4$ is four times larger than the fundamental domain for $\mathrm{SYM}(C_8) \approx D_8$, we have $[D_8 : \mathbb{Z}_4] = 4$. Of course, this index can simply be computed by dividing $|D_8|$ by 4, but that approach will not work when the groups involved are infinite.

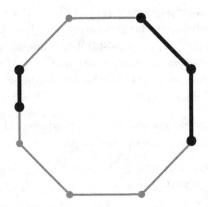

Fig. 1.25. A fundamental domain for the action of $D_8$ is highlighted on the left side of the $C_8$ graph, while a fundamental domain for the action of $\mathbb{Z}_4$ on $C_8$ is highlighted on the right.

The idea of finding fundamental domains for group actions, and using them to find generating sets, is much more general than what we have presented here. For example, instead of working with groups acting on graphs one could look at groups acting on the Euclidean plane. A simple example would be $\mathbb{Z} \oplus \mathbb{Z}$ where $(1,0)$ is a horizontal translation and $(0,1)$ is a vertical translation, both of unit length. The square $\mathcal{S}$ formed by taking the convex hull of the four vertices $(\pm 1/2, \pm 1/2)$ is a fundamental domain for this action. The set

$$S = \{g \in \mathbb{Z} \oplus \mathbb{Z} \mid g \neq (0,0) \text{ and } g \cdot \mathcal{S} \cap \mathcal{S} \neq \emptyset\}$$

is just $\{(\pm 1, 0), (0, \pm 1), (\pm 1, \pm 1)\}$. While this is larger than is necessary, it is indeed a generating set for $\mathbb{Z} \oplus \mathbb{Z}$.

We note that we used a variation of this technique when considering the symmetry group of a cube in Section 1.6.3. There we noted that a triangle formed a fundamental domain for the action of the symmetry group, and we took as our generating set the reflections in the sides of this triangle.

## 1.9 Words and Paths

The following set of definitions is our first entry into formal language theory. At the moment we are just establishing convenient terminology, but in Chapter 5 we explore some deeper connections between formal languages and infinite groups.

**Definition 1.60.** Given a set $S$, a finite sequence of elements from $S$, possibly with repetition, is called a *word*. The original set $S$ is the *alphabet* and the collection of all words, including the empty word consisting of a sequence of zero elements, is denoted $S^*$. This is the *free monoid* on $S$ and, while we will use this terminology, we shall not need to explore exactly what is implied by this term.

We let $S^{-1}$ denote a set of formal inverses to the elements of $S$. For example, if $S = \{a, b\}$ then $S^{-1} = \{a^{-1}, b^{-1}\}$. We can then form the free monoid $\{S \cup S^{-1}\}^*$ whose elements are finite lists of elements from $S$ and their formal inverses. We emphasize that these are formal inverses. If $S = \{a, b\}$ then $b, aba^{-1}, ab$, and $aba^{-1}a$ are all distinct elements in the free monoid $\{S \cup S^{-1}\}^*$.

We add two conventions: if $a \in S^{-1}$, so $a = x^{-1}$ for some $x \in S$, then we let $a^{-1}$ be an alternate notation for $x$. In other words, we declare $\left(x^{-1}\right)^{-1}$ to be $x$. We also allow ourselves to take formal inverses of words, not just elements. If

$$\omega = x_1 x_2 \cdots x_{k-1} x_k \in \{S \cup S^{-1}\}^*$$

then $\omega^{-1}$ stands for the word

$$\omega^{-1} = x_k^{-1} x_{k-1}^{-1} \cdots x_2^{-1} x_1^{-1} \in \{S \cup S^{-1}\}^*.$$

Given any word $\omega \in \{S \cup S^{-1}\}^*$, there is an associated edge path in the Cayley graph $\Gamma_{G,S}$. The path starts at the vertex corresponding to the identity and then traverses edges of $\Gamma_{G,S}$ as dictated by $\omega$. For example, consider $\mathbb{Z} \times \mathbb{Z}$ generated by $x = (1,0)$ and $y = (0,1)$. Then the word $\omega = xxy^{-1}x^{-1}yyyxxx$ describes the edge path illustrated in Figure 1.26.

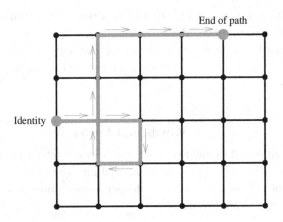

Fig. 1.26. The edge path $p_\omega$ associated to the word $\omega = xxy^{-1}x^{-1}yyyxxx$.

We will frequently refer to the edge path described by a word $\omega$, and we let $p_\omega$ denote this path. Thus given a word $\omega$ the path $p_\omega : [0,1] \to \Gamma_{G,S}$ will join the vertex corresponding to the identity to the vertex corresponding to the group element gotten by computing the product of $\omega$'s terms in $G$.

Conversely, every finite edge path in a Cayley graph $\Gamma_{G,S}$ describes a word in the generators and their inverses: one reads off the labels of the edges being traversed, and adds an inverse if they are travelling in the opposite direction of the orientation of the edge. Thus one sees that given a group $G$ and a finite generating set $S$ there is a one-to-one correspondence between

1. finite edge paths in the Cayley graph $\Gamma_{G,S}$ which begin at the vertex corresponding to the identity and
2. words in the free monoid $\{S \cup S^{-1}\}^*$.

Because of this connection between words and paths, the Cayley graph of a group $G$ is often thought of as a calculator for $G$. Let $g$ and $h$ be elements of $G$ and let $w_g$ and $w_h$ be words in $\{S \cup S^{-1}\}^*$ representing these elements. To compute the product $g \cdot h$, one can follow the edge path $p_g$ described by $w_g$ and then, starting at the vertex corresponding to $g$ instead of the vertex corresponding to the identity, trace out the edge path described by the word $w_h$. Continuing with the $\mathbb{Z} \times \mathbb{Z}$ example, if $g = y^2 x^3$ and $h = y^{-3}x$ then $g \cdot h = y^2 x^3 y^{-3} x$, as is shown in Figure 1.27. From this picture it is clear that $g \cdot h$ can be expressed as $x^4 y^{-1}$.

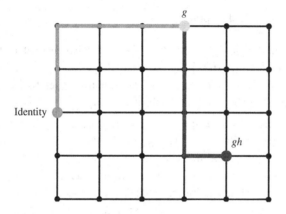

Fig. 1.27. Visualizing the product of $g = y^2 x^3$ and $h = y^{-3}x$.

## Exercises

(1) Enumerate all of the cycles in the graph of Figure 1.4.

(2) Construct infinitely many distinct biregular graphs where the white vertices have degree 2 and the black vertices have degree 3.

(3) Construct a graph $\Gamma$ where $\textsc{Sym}(\Gamma)$ is vertex transitive but not edge transitive.

(4) Construct a graph whose symmetry group is the cyclic group $\mathbb{Z}_n$ for each $n \geq 3$.

(5) There are regular graphs whose symmetry groups are edge transitive but not vertex transitive. (The smallest example is the "Folkman graph," pictures of which are easily found on the Internet.) Prove that any such graph must be bipartite. (A graph is *regular* if the valence of every vertex is the same.)

(6)  Let $\mathcal{P}_n$ be the graph formed by the edges of an $n$-prism (for $n \geq 3$). The graphs $\mathcal{P}_3, \mathcal{P}_4$ and $\mathcal{P}_5$ are shown in Figure 1.28. Prove that $\mathrm{SYM}(\mathcal{P}_n) \approx D_n \oplus \mathbb{Z}_2$ when $n \neq 4$. What goes wrong when $n = 4$?

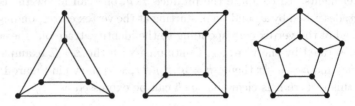

Fig. 1.28. The prism graphs referred to in Exercise 6.

(7)  Draw the Cayley graph of $G = \mathbb{Z}_5 \oplus \mathbb{Z}_2$ with respect to the generating set $\{(1,0), (0,1)\}$. What is the difference between this Cayley graph and the Cayley graph of $D_5$ (with respect to $\{\rho, f\}$)?

(8)  Let $\Gamma_{G,S}$ be the Cayley graph of a group $G$.

    a. Show by example that $\mathrm{SYM}(\Gamma_{G,S})$ can be a strictly larger group than $\mathrm{SYM}^+(\Gamma_{G,S}) \approx G$.

    b. Show by example that the size of the gap between $\mathrm{SYM}(\Gamma_{G,S})$ and $\mathrm{SYM}^+(\Gamma_{G,S})$ can be infinite. That is, find a group $G$ such that the index $[\mathrm{SYM}(\Gamma_{G,S}) : \mathrm{SYM}^+(\Gamma_{G,S})]$ is infinite.

(9)  Let $\Gamma$ be the Petersen graph. You may assume that $\mathrm{SYM}(\Gamma) \approx \mathrm{SYM}_5$. Prove that if $v$ is a vertex of $\Gamma$ then $|\mathrm{Stab}(v)| = 12$. Can you determine which group of order 12 this is?

(10)  Prove that a set of elements $S$ generates a group $G$ if and only if the only subgroup of $G$ that contains $S$ is $G$.

(11)  Prove that every finitely generated group is countable.

(12)  The first published Cayley graph, appearing in Cayley's original paper [Ca78], was the Cayley graph for $A_4$ with respect to the generating set $\{(123), (234)\}$. Draw this Cayley graph.

(13)  Let $G$ be the symmetry group of the regular octahedron.

    a. Compute the order of $G$.

    b. Find a set of generators $S$ for $G$ and explain how you know these form a set of generators. (Don't cheat and pick $S = G$!)

    c. Draw or explain in detail the shape and structure of the Cayley graph $\Gamma_{G,S}$.

(14)  What is the order of the symmetry group of a dodecahedron? Find a set of generators and describe the associated Cayley graph.

(15) Let $\Box^n$ be the $n$-cube formed by taking the convex hull of the points in $\mathbb{R}^n$ whose coordinates are $(\pm 1, \ldots, \pm 1)$. Let $\mathrm{Sym}(\Box^n)$ be the associated symmetry group.

    a. What is the order of $\mathrm{Sym}(\Box^n)$?

    b. Find a set of generators for $\mathrm{Sym}(\Box^n)$.

    c. Draw the Cayley graph of $\mathrm{Sym}(\Box^3)$ with respect to your chosen set of generators.

(16) The alternating group $A_4$ consists of the twelve even permutations of $\{1, 2, 3, 4\}$. The permutations (123) and (12)(34) generate $A_4$. Draw the associated Cayley graph.

(17) Let $G = \mathrm{SYM}(K_{3,3})$.

    a. Show that $G$ is vertex transitive.

    b. Show that $G$ is flag transitive.

    c. Show that $G$ is 2-transitive. That is, elements of $G$ can take any embedded arc of combinatorial length 2 to any other such arc.

    d. Show that $G$ is also 3-transitive.

    e. Is $G$ 4-transitive?

(18) Let $\Gamma$ be a graph and let $G \curvearrowright \Gamma$.

    a. If the action of $G$ on $\Gamma$ is vertex transitive, show that, for any $v$ and $w$ in $V(\Gamma)$, $\mathrm{Stab}(v) \approx \mathrm{Stab}(w)$.

    b. If you didn't already show it above, show that $\mathrm{Stab}(v)$ is conjugate to $\mathrm{Stab}(w)$.

(19) Given an edge $e$ in a graph $\Gamma$ there are two reasonable ways to define the *stabilizer of $e$*. One is to say

$$\mathrm{Stab}(e) = \{g \in G \mid g \cdot e = e\}$$

and the other is to say

$$\mathrm{Stab}'(e) = \{g \in G \mid g \text{ fixes } e \text{ pointwise}\}.$$

    a. Let $\Gamma$ be the wheel graph from Figure 1.22 and let $G = \mathrm{SYM}(\Gamma)$. Show that if $e$ is one of the edges attached to the central vertex, then

$$\mathrm{Stab}'(e) = \mathrm{Stab}(e) \approx \mathbb{Z}_2.$$

    b. Let $e$ be one of the exterior edges of the wheel graph $\Gamma$. Show that $\mathrm{Stab}'(e)$ is trivial while $\mathrm{Stab}(e) \approx \mathbb{Z}_2$.

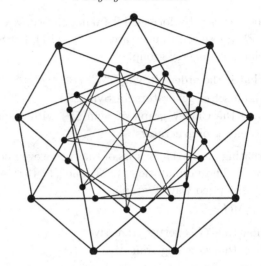

Fig. 1.29. The Doyle graph, which is vertex and edge transitive, but not flag transitive.

      c. Let $G \curvearrowright \Gamma$ be an arbitrary action of a group on a graph. Let $e$ be any edge of $\Gamma$. Prove that $\mathrm{Stab}'(e)$ is a subgroup of index at most 2 in $\mathrm{Stab}(e)$.

(20) Prove the following Corollary to Theorem 1.51: *For every finitely generated group $G$, there is a graph $\Gamma$, neither directed nor edge labelled, where* $\mathrm{SYM}(\Gamma) \approx G$.

(21) Let $H$ be the subset of permutations of the set of integers, $\mathrm{SYM}_{\mathbb{Z}}$, such that $h \in H$ if and only if there is a finite set of integers $C \subset \mathbb{Z}$ and a number $k$ such that $h(n) = n + k$ if $n \notin C$. In other words, the elements of $H$ all look like translations outside of a finite set of numbers. (Note: the subset $C$ and the number $k$ are both dependent on which element $h$ you are examining.)

      a. Show that $H$ is a subgroup of $\mathrm{SYM}_{\mathbb{Z}}$.
      b. Show that $H$ is finitely generated.

(The group $H$ is one example of an interesting family of groups called the *Houghton* groups.)

(22) Let $\Gamma$ be the Petersen graph, shown in Figure 1.8. The group of symmetries of $\Gamma$ is vertex transitive, just as is the case for a Cayley graph. Is there a group $G$ and a generating set $S$ such that the

Petersen graph is the underlying graph of the Cayley graph $\Gamma_{G,S}$? (Hint: The answer is no.)

(23) The graph $\Gamma$ shown in Figure 1.29 was discovered by Doyle in 1976 as part of his senior thesis (see [Ho81]). Prove that $\text{SYM}(\Gamma)$ is vertex transitive and edge transitive but it is not flag transitive.

# 2

# Groups Generated by Reflections

So, my interest in symmetry has not been misplaced.
–H. S. M. Coxeter (upon learning that his brain
displayed a high degree of bilateral symmetry)

Let $a$ and $b$ be reflections in parallel lines in the Euclidean plane (as in Figure 2.1). If we think of the line of reflection for $a$ as being $x = 0$ and the line of reflection for $b$ as being $x = 1$, then we may express $a$ as the function $a[(x, y)] = (-x, y)$ and $b$ as $b[(x, y)] = (2 - x, y)$. It follows that $ab[(x, y)] = (x - 2, y)$ and $ba[(x, y)] = (x + 2, y)$. Moreover, for any $n \in \mathbb{N}$, $(ab)^n$ is a horizontal translation to the left through a distance of $2n$ and $(ba)^n$ is a horizontal translation to the right through a distance of $2n$. Thus the reflections $a$ and $b$ generate an infinite group.

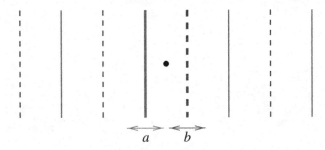

Fig. 2.1. The reflections $a$ (across the solid line) and $b$ (across the dashed line) generate an infinite group.

If the two lines of reflection actually met, at an angle of $\pi/n$, then the group generated by $a$ and $b$ would be the dihedral group of order $2n$, and the product $ab$ would be a rotation through an angle of $2\pi/n$. Thus the group we are considering is something like a dihedral group,

44

except that the two reflections generate a translation, not a rotation. As the group generated by two parallel reflections is infinite, it is referred to as the *infinite dihedral group*, denoted $D_\infty$.

Because $ab$ moves the line $x = 2$ to the line $x = 0$, which is fixed by $a$, and its inverse $ba$ returns the line $x = 0$ to the line $x = 2$, the reader can (and should!) verify that $ba \cdot a \cdot ab = bab$ is the reflection in the line $x = 2$. Similarly $aba$ gives a reflection in the line $x = -1$. Starting from these computations one can build an induction argument that proves that every vertical line $x = n$ is the axis of reflection for some element of $D_\infty$. In fact, these computations form the base cases of an induction argument (which we ask you to complete as Exercise 1) that establishes the following description of the elements of $D_\infty$:

**Proposition 2.1.** *Let $a$ be the reflection that fixes $x = 0$ and $b$ the reflection fixing $x = 1$, and let $D_\infty$ be the group generated by $a$ and $b$. Then every element of $D_\infty$ can be expressed as an alternating product of $a$'s and $b$'s. Further, if $n \in \mathbb{N}$, then the following hold.*

1. *The element $\underbrace{abab \cdots abab}_{2n \ letters} = (ab)^n$ is the horizontal translation described by $(x, y) \mapsto (x - 2n, y)$.*

2. *The element $\underbrace{baba \cdots baba}_{2n \ letters} = (ba)^n$ is the horizontal translation described by $(x, y) \mapsto (x + 2n, y)$.*

3. *The element $\underbrace{baba \cdots abab}_{2n - 1 \ letters} = (ba)^{n-1}b$ is the reflection that fixes $x = n$.*

4. *The element $\underbrace{abab \cdots baba}_{2n + 1 \ letters} = (ab)^n a$ is the reflection that fixes $x = -n$.*

The point in the middle of Figure 2.1 is not on an axis of reflection, hence, by the description given in Proposition 2.1, it is moved freely by $D_\infty$. This means we may apply the Drawing Trick (Remark 1.49) to construct the Cayley graph $\Gamma$ of $D_\infty$, with respect to $\{a, b\}$, using the orbits of this point as the vertices. Appealing to the convention of dropping double arrows when generators are of order 2, we get the Cayley graph embedded in $\mathbb{R}^2$ as in Figure 2.2. Looking at this Cayley graph one realizes that having $D_\infty$ act on the Euclidean plane is a bit of overkill: $D_\infty$ acts on the real line. The generator $a$ can be thought of as the reflection that fixes $0 \in \mathbb{R}$ and $b$ as the reflection fixing $1 \in \mathbb{R}$.

*Groups Generated by Reflections*

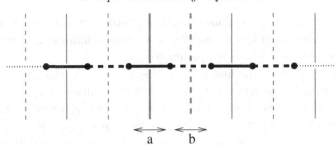

Fig. 2.2. The Cayley graph of the infinite dihedral group

The group $D_\infty$ is a non-abelian, infinite group that exhibits some striking behavior.

**Proposition 2.2.** *The subgroup $H$ of $D_\infty$ generated by $\{a, bab\}$ is a subgroup of index 2 that is isomorphic to $D_\infty$.*

*Proof.* Let $\alpha = a$ and $\beta = bab$. There is a homomorphism $\phi : H \to D_\infty$ defined by declaring $\phi(\alpha) = a$, $\phi(\beta) = b$ and then for an arbitrary element $\phi(x_1 x_2 \cdots x_n) = \phi(x_1)\phi(x_2) \cdots \phi(x_n)$, where $x_i \in \{\alpha, \beta\}$. This map is well defined, as it follows from the description in Proposition 2.1 that an element of $D_\infty$ has only one such expression as a product of $a$'s and $b$'s. That this map satisfies the condition $\phi(xy) = \phi(x)\phi(y)$, is immediate from its description. Similarly it is immediate that the homomorphism is surjective, as $a$ and $b$ are in the image, and $D_\infty$ is generated by $a$ and $b$.

To see that $\phi$ is injective, consider the possibility that some non-identity element in $H$ is taken to the identity in $D_\infty$. This element is either of even or odd length, and either it begins with $\alpha$ or $\beta$. But then the image under $\phi$ has the same form and, by Proposition 2.1, it is not the identity.

To see that $H$ is of index 2, we note that the elements of $H$ are simply the elements of $D_\infty$ that have an even number of $b$'s in their expression. Thus $D_\infty = H \cup bH$. $\qquad\square$

(The reader who finds Proposition 2.2 interesting should work through Exercise 3 at the end of this chapter.)

Having considered groups generated by two reflections, we now consider a few groups generated by three reflections. The Euclidean plane can be tiled by equilateral triangles. Fix a triangle $\mathbb{T}$ inside of such a tiling, and consider the group generated by the three reflections whose axes of reflection are the (extended) sides of $\mathbb{T}$. In Figure 2.3 we have

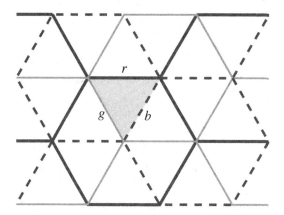

Fig. 2.3. A tiling of the Euclidean plane where the edges in the tiling have one of three types associated to them.

highlighted one such triangle; we refer to the reflections across the sides of this triangle as $r$, $g$ and $b$, as is indicated. As a quick exercise, the reader should verify that $r, rgr$ and $rbr$ are the reflections in the sides of the triangle immediately above $\mathbb{T}$ in this tiling. For example, in $rgr$ the rightmost "$r$" flips the triangle above $\mathbb{T}$ down to $\mathbb{T}$'s position; the middle "$g$" flips $\mathbb{T}$ to the left; then the leftmost $r$ flips $g \cdot \mathbb{T}$ up to the correct position.

Think of the edges of $\mathbb{T}$ as being wet with three types of paint, $r, b$ and $g$. When we apply the reflection $r$ to $\mathbb{T}$, the edges painted $b$ and $g$ are flipped upward, and they leave traces of their color on these edges in the tiling. If we continue to flip $\mathbb{T}$ across its edges, it will continue to leave traces of wet paint on the edges of the tiling. In this manner every edge in the tiling can be labelled as either $r, b$ or $g$. We have indicated this in Figure 2.3 with differing line styles. This labelling of the edges in the tiling makes it quite easy to determine the action of any particular element of $G$.

Let $\mathbb{T}'$ be any triangle in the tiling. Then $\mathbb{T}'$ can be joined to $\mathbb{T}$ by a *corridor* of triangles, that is, a sequence of triangles

$$\{\mathbb{T} = \mathbb{T}_0, \mathbb{T}_1, \mathbb{T}_2, \cdots, \mathbb{T}_n = \mathbb{T}'\}$$

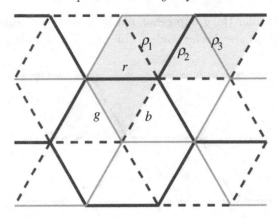

Fig. 2.4. A corridor joining two triangles in the tiling of the Euclidean plane.

where $\mathbb{T}_i$ intersects $\mathbb{T}_{i+1}$ in an edge (when $0 \leq i < n$). An example is shown in Figure 2.4. In this figure, if $\rho_i$ denotes a reflection across the associated edge, then the element $\rho_3 \rho_2 \rho_1 r$ moves $\mathbb{T}$ to $\mathbb{T}'$. As was noted previously, $\rho_1$ is a conjugate of $b$, namely $\rho_1 = rbr$. Thus we can rewrite $\rho_1 r$ as $rbr \cdot r = rb$. Since $\rho_1 r(= rb)$ takes $\mathbb{T}$ to the third triangle in the indicated corridor, it follows that $\rho_2 = \rho_1 r \cdot r \cdot r\rho_1 = rbrbr$. Hence

$$\rho_2 \rho_1 b = (rbrbr)(rbr)r = rbr.$$

Continuing in this fashion we get

$$\rho_3 \rho_2 \rho_1 b = rbrg$$

since $\rho_3 = \rho_2 \rho_1 r \cdot g \cdot r \rho_1 \rho_2 = rbr \cdot g \cdot rbr$.

**Lemma 2.3.** *Let $\mathbb{T}'$ be joined to $\mathbb{T}$ by a corridor*

$$\{\mathbb{T} = \mathbb{T}_0, \mathbb{T}_1, \mathbb{T}_2, \cdots, \mathbb{T}_n = \mathbb{T}'\}.$$

*Let $\ell_i \in \{r, g, b\}$ be the label of the edge $\mathbb{T}_{i-1} \cap \mathbb{T}_i$. Then the element $\ell_1 \ell_2 \cdots \ell_n$ takes $\mathbb{T}$ to $\mathbb{T}'$.*

*Proof.* The base case follows from the definition of $r, g$ and $b$.

Assuming our result is true for corridors of length $n$, we consider a corridor

$$\{\mathbb{T} = \mathbb{T}_0, \mathbb{T}_1, \mathbb{T}_2, \cdots, \mathbb{T}_n, \mathbb{T}_{n+1} = \mathbb{T}'\}.$$

Because $\rho_1 \rho_2 \cdots \rho_n$ takes $\mathbb{T}$ to $\mathbb{T}_n$, we know $\rho_n \rho_{n-1} \cdots \rho_1$ takes $\mathbb{T}_{n+1}$ to the triangle that is adjacent to $\mathbb{T}$ along the edge labelled $\rho_{n+1}$. Thus $\rho_{n+1} \rho_n \rho_{n-1} \cdots \rho_1$ takes $\mathbb{T}_{n+1}$ to $\mathbb{T}$, hence $\rho_1 \rho_2 \cdots \rho_n \rho_{n+1}$ takes $\mathbb{T}$ to $\mathbb{T}_{n+1}$. $\qquad \square$

**Remark 2.4.** Lemma 2.3 is an example of what we refer to as the "two wrongs make a right" principle. As we have frequently mentioned, our actions are always left actions. In particular, if we want to know where the element $brg$ takes $\mathbb{T}$, we start by applying $g$, then $r$, then $b$. And yet, Lemma 2.3 tells us we can work left-to-right, if we make one further mistake. Namely, after we apply $b$, we then apply "$r$" *thinking of this as the reflection across the edge labelled $r$ in $b \cdot \mathbb{T}$.* We then apply "$g$", repeating the same mistake, by reflecting $br \cdot \mathbb{T}$ across its edge labelled $g$.

This principle will be discussed again in Section 3.5.

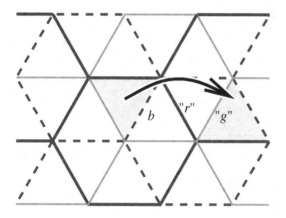

Fig. 2.5. The element $brg$ moves $\mathcal{T}$ to the triangle highlighted on the right.

Lemma 2.3 implies that $G$ acts transitively on the triangles of this tiling. That is, given any two triangles $\mathbb{T}$ and $\mathbb{T}'$, there is an element $h \in G$ such that $h \cdot \mathbb{T} = \mathbb{T}'$. Further, let $h = \rho_1 \rho_2 \rho_3 \cdots \rho_n$ where each $\rho_i \in \{r, g, b\}$. Then there is a corridor $\{\mathbb{T} = \mathbb{T}_0, \mathbb{T}_1, \mathbb{T}_2, \cdots, \mathbb{T}_n\}$ where the label of the edge $\mathbb{T}_{i-1} \cap \mathbb{T}_i$ is $\rho_i$. By Lemma 2.3, $h \cdot \mathbb{T} = \mathbb{T}_n$.

There is a theorem of Euclidean geometry that says any isometry of the plane is determined by where it sends three non-collinear points. Using this one can see that the action of $G$ on the triangles of this tiling is simply transitive. If there is an $h \in G$ with $h \cdot \mathbb{T} = \mathbb{T}$ then, by the theorem stated above, $h$ is the identity. Since $G$ acts simply transitively on the triangles, $G$ also acts simply transitively on the barycenters of the triangles in the tiling. We may then apply the Drawing Trick using the barycenter, $x$, of our chosen triangle $\mathbb{T}$. The barycenters of the triangles form the orbit of $x$ under the actions of $G$, hence they correspond to

the vertices of the Cayley graph. When the edges of the Cayley graph
are added, one sees that the Cayley graph is the vertices and edges of
the regular hexagonal tiling of plane that is dual to the original tiling
by equilateral triangles. This is shown in Figure 2.6.

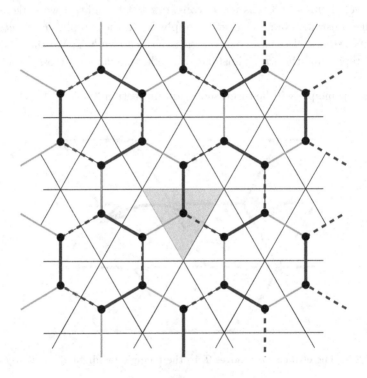

Fig. 2.6. The Cayley graph of the reflection group $W_{333}$.

There are two other tilings of the Euclidean plane that lead naturally
to infinite groups generated by three reflections. One can tile the plane
by triangles with interior angles $\{\pi/2, \pi/4, \pi/4\}$ and one can also tile
the plane by triangles with interior angles $\{\pi/2, \pi/3, \pi/6\}$. The groups
generated by reflecting in the sides of such triangles are similar to the
one we have discussed above, but they have their individual quirks.

At this point we have a reasonably long list of "groups generated by
reflections." There are the finite dihedral groups as well as $D_\infty$ and the
groups generated by reflections in the sides of certain Euclidean triangles.
The symmetry groups of all five Platonic solids are also groups generated
by reflections, as are the symmetry groups of cubes in all dimensions.
There is a well-developed theory that contains these groups and many

others; the general class of groups being considered was identified by Jacques Tits who called them *Coxeter groups* in honor of Coxeter's work on symmetry groups of regular polygons. Within the general theory of Coxeter groups, the groups we have been considering in this chapter are called "Coxeter groups of type __": the infinite dihedral group $D_\infty$ is the Coxeter group of type $\widetilde{A}_1$; the group generated by reflections in the sides of an equilateral triangle is of type $\widetilde{A}_2$; the group generated in the sides of a triangle with angles $\{\pi/2, \pi/4, \pi/4\}$ is of type $\widetilde{B}_2$; and the group generated by reflections in the sides of a triangle with angles $\{\pi/2, \pi/3, \pi/6\}$ has type $\widetilde{G}_2$.

This notation makes sense in the context of the classification of finite Coxeter groups and Coxeter groups generated by reflections in Euclidean spaces (of all dimensions), but it is not of much use to us. We will continue to refer to $D_\infty$ as $D_\infty$ and we will refer to the group generated by reflections in the sides of a Euclidean triangle with interior angles $\{\pi/p, \pi/q, \pi/r\}$ as $W_{pqr}$. (The letter $W$ is commonly used to denote a Coxeter group.)

**Example 2.5.** Theorem 1.58 was written in terms of fundamental domains for actions on graphs, but it did not need to be so restrictive. In Figure 2.7 we show the axes of reflection for elements of $W_{333}$ and have highlighted three such axes that form an equilaterial triangle, which is four times bigger than the fundamental domain for $W_{333} \curvearrowright \mathbb{R}^2$. If $H$ is the subgroup generated by these three reflections, then $H \approx W_{333}$ is a group generated by reflections in the sides of an equilateral triangle. Using the argument of Theorem 1.58 in this context we find that $H$ has index 4 in the original $W_{333}$.

The reader who is interested in Coxeter groups can learn the basics by reading [Hu90]. An excellent book introducing the deep combinatorial information associated to Coxeter groups should read [BjBr05]. A reader with a stronger background in mathematics than we are assuming in this text, who is interested in the geometry and topology of Coxeter groups, should read [Da08].

### Exercises

(1)    Prove Proposition 2.1. This is perhaps easiest to do if you view $a$ as the function $a(x) = -x$ and $b$ as the function $b(x) = 2 - x$, remembering that the group operation is function composition.

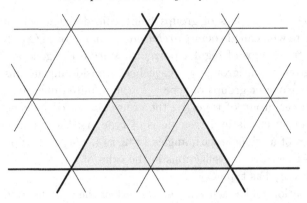

Fig. 2.7. The group $W_{333}$ contains an index 4 subgroup that is isomorphic to $W_{333}$.

(2)  Recall that $T_{2,2}$ is the regular 2-valent, bipartite tree. Prove that $D_\infty \approx \text{Sym}^+(T_{2,2})$.

(3)  Let the two reflections generating $D_\infty$ be denoted $a$ and $b$, and use the same conventions for the action $D_\infty \curvearrowright \mathbb{R}$ as in Proposition 2.1.

    a.  Prove that the subgroup $\langle a, babab \rangle$ is isomorphic to $D_\infty$ and is of index 3 in the original group.

    b.  Prove that for any $n \geq 1$ the group $D_\infty$ contains a subgroup that is isomorphic to $D_\infty$ and of index $n$.

    c.  Prove that if $\alpha$ is any automorphism of $D_\infty$ then $\alpha$ takes $a$ and $b$ to reflections fixing $x = n$ and $x = n + 1$, for some $n \in \mathbb{Z}$.

    d.  Prove that the center of $D_\infty = \{e\}$ and conclude that $D_\infty \approx \text{Inn}(D_\infty) < \text{Aut}(D_\infty)$.

    e.  Show that the index of $\text{Inn}(D_\infty)$ in $\text{Aut}(D_\infty)$ is 2. (Hint: There is an automorphism given by $\phi(a) = b$ and $\phi(b) = a$, which is not an inner automorphism.)

    f.  Prove that $\text{Aut}(D_\infty) \approx D_\infty$.

(4)  Determine how the element $rgb$ acts on the tiling of the Euclidean plane by regular triangles. What about $gbr$? or $brg$?

(5)  Tile the Euclidean plane by squares of side length 1. Let $W$ be the group generated by the four reflections in the (extended) sides of any one square. Draw the Cayley graph of $W$ and prove that $W \approx D_\infty \oplus D_\infty$.

(6)  Prove that $\mathbb{Z} \oplus \mathbb{Z}$ is a subgroup of $W_{333}$.

(7)  Draw the Cayley graphs of $W_{244}$ and $W_{236}$.

(8) Show that $W_{236}$ contains a finite index subgroup isomorphic to $W_{333}$.

(9) Show that each of the groups $W_{236}, W_{244}$, and $W_{333}$ contains a proper subgroup isomorphic to itself.

**Bonus:** For which values of $n$ is $W_{333}$ a subgroup of index $n$ in itself?

# 3

# Groups Acting on Trees

Eine abstrakte Theorie der diskontinuierlichen Gruppen, die von vornherein darauf abzielt, *unendliche* Gruppen einzubegreifen, wird in dem Studium der freien Gruppen irhren natürlichen Ausgangspunkt finden.

–Jakob Nielsen

## 3.1 Free Groups

Free groups are central to the study of infinite groups and provide relatively accessible examples that highlight the power of mixing algebraic and geometric approaches. Free groups can be defined in a number of ways. The standard method, which we give in Section 3.1.1, is formal and algebraic. Another common method is to use actions on trees, which is the perspective we develop in Section 3.4. While the formal definition lacks the intuition building power of the geometric perspective, it is often the easiest definition to use when trying to prove that a given group is a free group.

### 3.1.1 Free Groups of Rank n

**Definition 3.1.** Let $S = \{x_1, \dots, x_n\}$ be a set of elements in a group, $G$. A word $\omega \in \{S \cup S^{-1}\}^*$ is said to be *freely reduced* if it does not contain a subword consisting of an element adjacent to its formal inverse. For example, the word $\omega = xyx^{-1}y^{-1}$ is freely reduced while $\omega' = xy^{-1}yxy$ is not. The group $G$ is a *free group with basis $S$* if $S$ is a set of generators for $G$ and no freely reduced word in the $x_i$ and their inverses represents the identity.

The *rank* of a free group with basis $S$ is the number of elements in $S$. We denote a free group of rank $n$ by $\mathbb{F}_n$.

In the definition above we explicitly assume that $S$ is a finite set, but one can discuss free groups of infinite rank, where the generating set $S$ is not finite. Before encountering such groups, we establish some important results about free groups of finite rank, not the least of which is that they exist!

**Theorem 3.2.** *Given any $n \in \mathbb{N}$, there is a free group of rank $n$.*

We only sketch the argument. A very similar argument is given in Section 3.6, where the situation is a bit more complicated.

*Outline of proof.* Let $S = \{x_1, \ldots, x_n\}$ be $n$ distinct symbols. Consider the equivalence relation on all words, in $\{S \cup S^{-1}\}^*$ induced by

$$a_1 \cdots a_{i-1} a_i a_i^{-1} a_{i+1} \cdots a_k \sim a_1 \cdots a_{i-1} a_{i+1} \cdots a_k$$

where we have cancelled an adjacent pair of elements because they are inverses of each other.

The elements of the group $\mathbb{F}_n$ are the equivalence classes of words in $\{S \cup S^{-1}\}^*$ under the equivalence relation $\sim$. If $\omega$ is a given word, we denote its equivalence class as $[\omega]$ and claim that the following binary operation defines a group structure on this set: $[\alpha][\beta] = [\alpha\beta]$ where $\alpha\beta$ is just the concatenation of the two words $\alpha$ and $\beta$. The identity element is then $[\varepsilon]$, where $\varepsilon$ represents the empty string, and the inverse of $[\omega]$ is $[\omega^{-1}]$ where $\omega^{-1}$ is the formal inverse of the word $\omega$ (as described in Section 1.9). $\qquad \square$

One should think of the elements of $\mathbb{F}_n$ as corresponding to freely reduced words, and the following exercise justifies this intuition.

**Exercise 3.3.** Show that every equivalence class contains exactly one freely reduced word from $\{x_1, \ldots, x_n, x_1^{-1}, \ldots, x_n^{-1}\}^*$.

The reason for introducing the equivalence relation, and not simply stating that freely reduced words are the elements of $\mathbb{F}_n$, is that the concatenation of two freely reduced words is not necessarily freely reduced.

### 3.1.2 $\mathbb{F}_2$ *as a Group of Tree Symmetries*

As will become more clear as this chapter progresses, there is a direct connection between free groups and free actions on trees. One way to produce a free action of a given group $G$ on a graph is via Cayley's Better Theorem (Theorem 1.42), which we exploit in the following proposition.

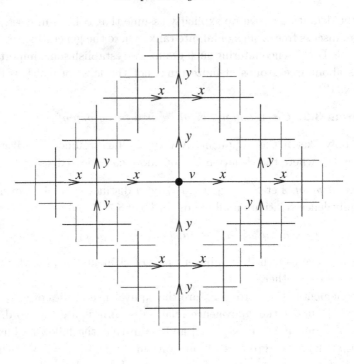

Fig. 3.1. The Cayley graph of $\mathbb{F}_2$.

**Proposition 3.4.** *The Cayley graph of $\mathbb{F}_2$ with respect to $\{x, y\}$ is an oriented version of $\mathcal{T}_4$, the uniformly 4-valent tree, shown in Figure 3.1.*

*Proof.* As $\mathbb{F}_2$ is generated by two elements, the associated Cayley graph will be uniformly 4-valent. Appealing to the correspondence between words and paths, we note that because no freely reduced word represents the identity, this Cayley graph of $\mathbb{F}_2$ contains no cycles. Thus the Cayley graph is a connected, uniformly 4-valent tree, $\mathcal{T}_4$.                    □

Since every group acts on its Cayley graph (Theorem 1.51) we immediately get:

**Corollary 3.5.** *The free group $\mathbb{F}_2$ can be viewed as a group of symmetries of $\mathcal{T}_4$.*

The group $\mathbb{F}_2$ acts on $\mathcal{T}_4$ on the left. So, for example, the generator $x$ takes the vertex corresponding to a freely reduced word $\omega$ to the vertex corresponding to the freely reduced word equivalent to $x \cdot \omega$. One can write a formula for this action. Let $\omega$ be a freely reduced word, and let $\omega = \alpha \cdot \omega'$, where $\alpha$ is a single letter. Then the action of $x$ is given by

$$x \cdot \omega = x \cdot \alpha \cdot \omega' = \begin{cases} x\omega & \text{if } x^{-1} \neq \alpha \\ \omega' & \text{if } x^{-1} = \alpha. \end{cases}$$

The action of $x \in \mathbb{F}_2$ on $\mathcal{T}_4$ is something of a "shift to the right." To see this, consider the subgraph of $\mathcal{T}_4$ induced by the vertices corresponding to $\{x^n \mid n \in \mathbb{Z}\}$. Left multiplication by $x$ shifts this combinatorial line one unit to the right, thereby shifting the entire Cayley graph, as is illustrated in Figure 3.2. The action of $y$ is similar, the main difference being that the combinatorial line induced by vertices corresponding to $\{y^n \mid n \in \mathbb{Z}\}$ is shifted one unit up.

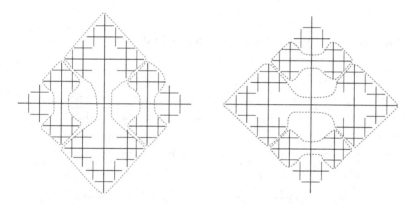

Fig. 3.2. Left multiplying by $x$ moves the vertical florets to the right, as in the left-hand side of the picture; left multiplying by $y$ moves the horizontal florets up.

This is perhaps a good time for a reminder: visualizing the action of a given element in a group $G$ on its Cayley graph is not the same as following paths from the identity to that element in a Cayley graph. Consider the specific example shown in Figure 3.3. Here we have an edge path between the vertex associated with the identity, $v$, and another vertex $w$, in the Cayley graph of $\mathbb{F}_2$. The edge path is labelled by the word $xy^{-1}x^{-1}y$. One might guess that the edge path traces out the motion of $v$ under the action of these four transformations, but that is not the case.

In Figure 3.4 we see the movement of the vertex $v$ under the action of $xy^{-1}x^{-1}y$. Remember that the action of $\mathbb{F}_2$ is on the left, so that the first thing one does is apply $y$ to $v$, moving it up. Then one applies $x^{-1}$

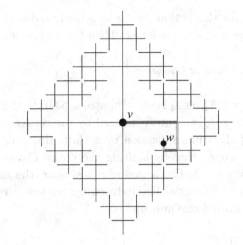

Fig. 3.3. The edge path connecting $v$ and $w$ in the Cayley graph of $\mathbb{F}_2$.

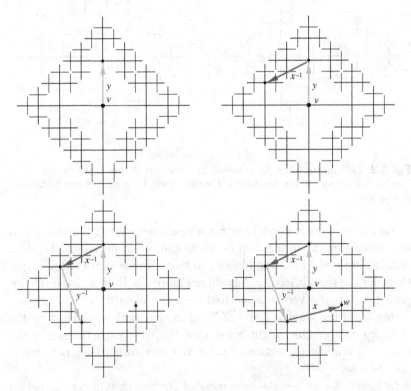

Fig. 3.4. Getting from $v$ to $w$ via the action of $\mathbb{F}_2$.

to the vertex $y \cdot v$, as shown in the upper right-hand corner. Then $y^{-1}$ moves $x^{-1}y \cdot v$ down, and $x$ moves $y^{-1}x^{-1}y \cdot v$ to the vertex $w$. Thus while the end result is the same, the route that the vertex $v$ travels is not along the path associated to the word $xy^{-1}x^{-1}y$. (This comment is similar to the discussion around Lemma 2.3.)

**Remark 3.6.** The reader should keep in mind that free groups and free *abelian* groups are different. Free abelian groups are the groups of the form $\mathbb{Z} \oplus \mathbb{Z} \oplus \mathbb{Z} \oplus \cdots$; they are not the same as the free groups $\mathbb{F}_n$. In particular, free abelian groups are commutative, while free groups are not.

### 3.1.3 Free Groups in Nature

The definition of free groups in terms of freely reduced words may appear to be quite formal, and in fact free groups may appear to have been created out of whole cloth, but these groups do arise in a startling number of mathematical contexts. (This is what mathematicians mean when they say something arises "naturally" or "in nature.") If you have studied non-Euclidean geometry, you can find free groups inside of the group of isometries of the hyperbolic plane (Exercise 5). Directly related to that situation, consider the group $\mathrm{SL}_2(\mathbb{Z})$ consisting of two-by-two matrices with integer entries and determinant 1. The subgroup of $\mathrm{SL}_2(\mathbb{Z})$ generated by

$$S = \left\{ \begin{bmatrix} 1 & 0 \\ 2 & 1 \end{bmatrix}, \begin{bmatrix} 1 & 2 \\ 0 & 1 \end{bmatrix} \right\}$$

has index 12 in $\mathrm{SL}_2(\mathbb{Z})$. In fact, this subgroup is the kernel of the map $\phi : \mathrm{SL}_2(\mathbb{Z}) \to \mathrm{SL}_2(\mathbb{Z}_2)$. (It is clear that $S \subset \ker(\phi)$ but it is not obvious that $S$ generates this kernel. Proving this fact would be a challenging exercise.)

**Proposition 3.7.** *The subgroup of* $\mathrm{SL}_2(\mathbb{Z})$ *generated by*

$$S = \left\{ \begin{bmatrix} 1 & 0 \\ 2 & 1 \end{bmatrix}, \begin{bmatrix} 1 & 2 \\ 0 & 1 \end{bmatrix} \right\}$$

*is isomorphic to* $\mathbb{F}_2$.

*Proof.* Let $l = \begin{bmatrix} 1 & 0 \\ 2 & 1 \end{bmatrix}$ and $r = \begin{bmatrix} 1 & 2 \\ 0 & 1 \end{bmatrix}$. We need to show that no freely reduced word in $\{l, r, l^{-1}, r^{-1}\}^*$ represents the identity. We do this using the action of $\mathrm{SL}_2(\mathbb{Z})$ on the Euclidean plane. We think of the point $(a, b)$ as the tip of the vector $\begin{bmatrix} a \\ b \end{bmatrix}$, and allow the elements of our subgroup to act on points in the plane via matrix multiplication. For example, the generator $l$ acts as follows:

$$\begin{bmatrix} 1 & 0 \\ 2 & 1 \end{bmatrix} \cdot \begin{bmatrix} a \\ b \end{bmatrix} = \begin{bmatrix} a \\ 2a + b \end{bmatrix}.$$

Thus $l \cdot (a, b) = (a, 2a + b)$. A similar computation establishes that $r \cdot (a, b) = (a + 2b, b)$. Inverting the matrices gives $l^{-1} \cdot (a, b) = (a, -2a + b)$ and $r^{-1} \cdot (a, b) = (a - 2b, b)$.

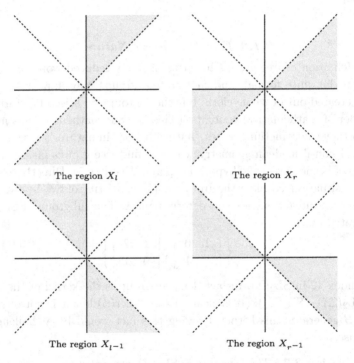

The region $X_l$

The region $X_r$

The region $X_{l^{-1}}$

The region $X_{r^{-1}}$

Fig. 3.5. Four subsets in the plane corresponding to the elements of $\{l, r, l^{-1}, r^{-1}\}$.

In Figure 3.5 we show specific subsets of the plane, each affiliated with a generator or an inverse of a generator. These subsets are unions of open sectors, each of whose boundary is contained in the union of the coordinate axes and the lines $y = x$ and $y = -x$. For example, $X_l$ is the union of two open sectors, where the bounding edges are $x = 0$ and $y = x$. (The point $(1, 3) \in X_l$, while $(1, 1) \notin X_l$, for example.) The subset $X_{l^{-1}}$ is the reflection of $X_l$ across the $y$-axis. Similarly, $X_r$ is the union of two open sectors, where the bounding edges are $y = 0$ and $y = x$; $X_{r^{-1}}$ is the reflection of $X_r$ across the $x$-axis.

Because $l$ is a linear transformation, where we know $l \cdot (0,1) = (0,1)$ and $l \cdot (1,1) = (1,3)$, we see that $l \cdot X_l$ is properly contained inside of $X_l$. Similarly, since $l \cdot (1,0) = (1,2)$ and $l \cdot (1,1) = (1,3)$, we see that $l \cdot X_r \subset X_l$. One more computation shows that $l \cdot X_{r^{-1}}$ is also contained in $X_l$. In fact, this containment property holds for each $w \in \{l, r, l^{-1}, r^{-1}\}$. That is, if $w \in \{l, r, l^{-1}, r^{-1}\}$ then

$$w \cdot X_y \subset X_w \text{ for } y \neq w^{-1},$$

where the containment is always proper.

Let $\omega = w_1 w_2 \cdots w_n$ be a freely reduced word in $\{l, r, l^{-1}, r^{-1}\}^*$. In order to prove that $\{l, r\}$ generates a free group of rank 2, we need to show that $\omega$ is not the identity. In other words, we need to show that $\omega$ does not act trivially on the plane. If $|\omega| = 1$, then it is clear from the definition that $\omega \neq e$. Furthermore, because $\omega$ is freely reduced, we may apply the containment property described above, one letter at a time:

$$\omega \cdot X_{w_n} = w_1 w_2 \cdots w_{n-2} w_{n-1} w_n \cdot X_{w_n}$$
$$\subset w_1 w_2 \cdots w_{n-2} w_{n-1} \cdot X_{w_n} \subset w_1 w_2 \cdots w_{n-2} \cdot X_{w_{n-1}}$$
$$\subset \cdots \subset w_1 X_{w_2} \subset X_{w_1}$$

where each of the containments is proper. Thus $\omega \cdot X_{w_n}$ is a proper subset of $X_{w_1}$. It follows that $\omega \cdot X_{w_n} \neq X_{w_n}$ and therefore $\omega \neq e$. $\square$

**Remark 3.8.** The fact that $SL_2(\mathbb{Z})$ contains a free subgroup is not an isolated fact. There is a celebrated theorem of Jacques Tits [Ti72] that says: *Let $G$ be a finitely generated group that can be faithfully represented as a group of matrices.*[1] *Then either $G$ contains a finite index subgroup that is solvable, or $G$ contains a subgroup isomorphic to $\mathbb{F}_2$.* The reader unfamiliar with solvable groups can interpret this result as saying: linear groups contain free subgroups unless they obviously do not.

In the example above we demonstrated that $\mathbb{F}_2 < SL_2(\mathbb{Z})$, and in Remark 3.8 we indicated that free groups commonly arise as subgroups of linear groups. Free groups also arise as subgroups in other important classes of groups. For example, a continuous, one-to-one and onto map $\phi : \mathbb{R} \rightarrow \mathbb{R}$, whose inverse is also continuous, is called a *homeomorphism*

---

[1] A group that can be faithfully represented as a matrix group is called a *linear group*.

of $\mathbb{R}$. Examples of homeomorphisms are $f(x) = x^3$, $f(x) = 7 - x$ and $f(x) = x + \sin(x)$. The collection of all such continuous maps forms a very large group, Homeo($\mathbb{R}$), whose binary operation is function composition. The identity element is the function $f(x) = x$, and inverses exist because we have restricted ourselves to functions that have inverses.

A number of interesting infinite groups arise as subgroups of Homeo($\mathbb{R}$). For example, the subgroup generated by $f(x) = x^3$ is infinite cyclic. The subgroup $H < $ Homeo($\mathbb{R}$) generated by $f(x) = x^3$ and $g(x) = x^5$ is a free abelian group of rank 2. To prove this, note that any $h \in H$ has the form $h(x) = x^{3^m 5^n}$. Thus there is a map from $H$ to $\mathbb{Z} \oplus \mathbb{Z}$ defined by $\phi : h(x) \to (m, n)$ when $h(x) = x^{3^m 5^n}$. Checking that $\phi$ is an isomorphism is a straightforward exercise. The infinite dihedral group, $D_\infty$, is an example of a non-abelian, infinite subgroup of Homeo($\mathbb{R}$).

We construct an example of a free subgroup of Homeo($\mathbb{R}$) using two explicit functions. Let $f(x)$ be the function whose graph is shown in Figure 3.6. This function is defined on $[0, 1]$ by

$$\hat{f}(x) = \begin{cases} 4x & \text{if } 0 \le x \le \frac{1}{5} \\ \frac{4}{5} + \frac{1}{4}\left(x - \frac{1}{5}\right) & \text{if } \frac{1}{5} \le x \le 1 \end{cases}$$

and is extended to the real line by $f(x) = \lfloor x \rfloor + \hat{f}(x - \lfloor x \rfloor)$. Define $X_f$ to be small closed intervals around the integers,

$$X_f = \bigcup_{i \in \mathbb{Z}} \left[ i - \frac{1}{5}, i + \frac{1}{5} \right].$$

This subset of $\mathbb{R}$ is indicated in both the domain and range in Figure 3.6. In this picture you can see that $f$ takes $(1/5, 4/5)$ into $(4/5, 1)$. In general,

$$f\left[\left(i + \frac{1}{5}, i + \frac{4}{5}\right)\right] \subset \left(i + \frac{4}{5}, i + 1\right) \subset X_f,$$

for all $i \in \mathbb{Z}$. Thus $f$ takes points that are outside of $X_f$, into $X_f$.

The argument above provides the base case for an induction argument establishing that, for any positive integer $n$, if $x \notin X_f$ then $f^n(x) \in X_f$. An important fact to notice is that if $x \notin X_f$ then $f(x) \in (i - 1/5, i)$ for some $i \in \mathbb{Z}$. But $f[(i - 1/5, i)] \subset (i - 1/5, i)$, so $f^2(x)$ is also in $(i - 1/5, i)$. For the induction step we may assume that if $x \notin X_f$ then $f^{n-1}(x) \in (i - 1/5, i)$ for some $i \in \mathbb{Z}$, and we then simply note that $f^n(x) = f\left(f^{n-1}(x)\right)$ must also be in $(i - 1/5, i) \subset X_f$.

The graph of the inverse of $f(x)$ is just the reflection of the graph of $f(x)$ across the line $y = x$. Thus the graph of $f^{-1}(x)$ is also contained

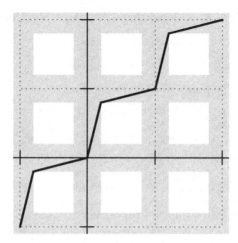

Fig. 3.6. The graph of the function $f$, which takes points in the open intervals $\{i + (1/5, 4/5) \mid i \in \mathbb{Z}\}$ to points in the open intervals $\{i + (4/5, 1) \mid i \in \mathbb{Z}\}$.

in the grid formed by $X_f$ in the domain and in the range. In this case, if $x \in (i - 4/5, i - 1/5)$ then $f^{-1}(x) \in (i - 1, i - 4/5)$, so $f^{-1}(x) \in X_f$. Further, since $f^{-1}$ takes $(i - 1, i - 4/5)$ into $(i - 1, i - 4/5)$, $f^{-n}(x) \in X_f$ for any $n \in \mathbb{N}$ and $x \notin X_f$.

Our second function is just a shifted copy of $f$. Define $g(x)$ to be $f(x - 1/2) + 1/2$ and let

$$X_g = \bigcup_{i \in \mathbb{Z}} [i + 1/2 - 1/5, i + 1/2 + 1/5].$$

Then the same arguments as are given above show that, for any non-zero integer $n$, positive or negative, $g^n(x) \in X_g$ if $x \notin X_g$.

We can now establish our claim that the subgroup of $\mathrm{Homeo}(\mathbb{R})$ generated by $f$ and $g$ is a free group, using an argument that is similar to the proof of Proposition 3.7.

**Proposition 3.9.** *The group of homeomorphisms of the real line generated by $f$ and $g$ (defined above) is a free group of rank 2.*

*Proof.* Let $\omega = w_1 \cdots w_n$ be a freely reduced word in $\{f, g, f^{-1}, g^{-1}\}^*$. We show that $\omega \neq e$ by showing that $\omega$ does not act trivially on $\mathbb{R}$. By collecting terms we may express $\omega$ as $w_1^{n_1} w_2^{n_2} \cdots w_k^{n_k}$ where $w_i \neq w_{i+1}$ or $w_{i+1}^{-1}$.

Notice that $X_f \cap X_g = \emptyset$ and that $1/4 \notin X_f \cup X_g$. Define

$$X_i = \begin{cases} X_f & \text{if } w_i \in \{f, f^{-1}\} \\ X_g & \text{if } w_i \in \{g, g^{-1}\}. \end{cases}$$

If we apply $\omega$ to the point $x = 1/4$ we see that

$$\omega(1/4) = w_1^{n_1} w_2^{n_2} \cdots w_{k-1}^{n_{k-1}} w_k^{n_k}(1/4)$$
$$\subset w_1^{n_1} w_2^{n_2} \cdots w_{k-1}^{n_{k-1}}(X_k) \subset w_1^{n_1} w_2^{n_2} \cdots w_{k-2}^{n_{k-2}}(X_{k-1})$$
$$\subset \cdots \subset w_1^{n_1} w_2^{n_2}(X_3) \subset w_1^{n_1}(X_2) \subset X_1 \subset X_f \cup X_g.$$

But $1/4 \notin X_f \cup X_g$ so $\omega(1/4) \neq 1/4$ and therefore $\omega \neq e$.  $\square$

The method used to establish Propositions 3.7 and 3.9 is often referred to as the Ping Pong Lemma. The version of the Ping Pong Lemma that was used to establish 3.9 is stated as:

**Lemma 3.10** (Ping Pong Lemma). *Let $G$ act on $X$ and let $\mathcal{S} = S \cup S^{-1}$ be a set of generators and their inverses. For each $s \in S$ let $X_s$ be a subset of $X$, and let $p$ be a point in $X \setminus \{\cup_{s \in S} X_s\}$. If*

1. *$s(p) \in X_s$ for each $s \in S$; and*
2. *$s(X_t)$ is a proper subset of $X_s$, for each $t \in \mathcal{S} \setminus \{s^{-1}\}$,*

*then $G$ is a free group with basis $S$.*

As we have essentially proved the Ping Pong Lemma in two specific cases, we leave the proof of the general result as an end of chapter exercise.

**Remark 3.11.** Proposition 3.9 was originally presented in a paper by Curtis Bennett [Be97]. The functions $f$ and $g$ used in Proposition 3.9 are somewhat complicated. There are, in fact, more elementary functions inside of Homeo($\mathbb{R}$) that generate free subgroups. For example, if $p$ is an odd prime, then the functions $f(x) = x^p$ and $g(x) = x + 1$ generate a free subgroup of Homeo($\mathbb{R}$). This seemingly accessible result is strikingly difficult to prove (see [Wh88]).

Our examples above are by no means an exhaustive list of places where free groups appear in mathematics. For example, free groups are an integral part of the proof of the Banach–Tarski Paradox.[2] If you are

---

[2] In 1924, Banach and Tarski proved that a solid ball in $\mathbb{R}^3$ can be decomposed into a finite number of pieces, which can then be reassembled to form two disjoint balls, each of the same volume as the single ball you started with. See [Wa86] for more information.

interested in the diverse places where free groups arise, [Gl92] is a nice expository article on this topic.

## 3.2 $\mathbb{F}_3$ is a Subgroup of $\mathbb{F}_2$

It is easy to see that a free group of rank 2 sits as a subgroup inside a free group of rank 3 (or rank 4, 5, ...). If $\{x_1, x_2, \ldots, x_n\}$ is a basis for $\mathbb{F}_n$, then no freely reduced word in $\{x_1, x_2, x_1^{-1}, x_2^{-1}\}^*$ can equal the identity. Thus the subgroup generated by $\{x_1, x_2\}$ is a free group of rank 2. On the other hand, the following result is surprising.

**Proposition 3.12.** *There is a finite index subgroup of $\mathbb{F}_2$ that is a free group of rank 3.*

In the remainder of this section we outline a proof of this proposition. Two key steps are left as exercises that the reader should complete. Our argument exploits the fact that elements of $\mathbb{F}_2$ can be assigned lengths. Since every $g \in \mathbb{F}_2$ can be uniquely expressed as a freely reduced word in the generators and their inverses, we define the *length* of $g$ to be the number of terms in this freely reduced expression. We denote this number by $|g|$. As an example,

$$|x^3 x^{-1} y^2 xy^{-5} y| = |x^2 y^2 xy^{-4}| = 9.$$

This notion of length has the useful property that $|g| = |g^{-1}|$, since the freely reduced expression for $g^{-1}$ is the formal inverse of the freely reduced expression for $g$.

Let $H$ be the subset of $\mathbb{F}_2$ consisting of elements of even length:

$$H = \{g \in F_2 \mid |g| \text{ is even}\}.$$

Since $|g| = |g^{-1}|$, this set is closed under taking inverses. Because cancellation in non-freely reduced words occurs in pairs, if $|a|$ and $|b|$ are both even, $|ab|$ will also be even, although it could be less than $|a| + |b|$. Thus $H$ is closed under products and inverses, and therefore it is a subgroup of $\mathbb{F}_2$. We call $H$ the *even subgroup* of $\mathbb{F}_2$. Finally, because the length of every element in $\mathbb{F}_2$ is either even or odd, $[\mathbb{F}_2 : H] = 2$.

**Exercise 3.13.** Prove that the even subgroup of $\mathbb{F}_2$ is generated by $S = \{x^2, xy, xy^{-1}\}$. (Hint: Start by proving that every freely reduced word of length 2 can be expressed as a product of elements of $S \cup S^{-1}$, and then proceed by induction.)

Let $a = x^2, b = xy$ and $c = xy^{-1}$, and let $\omega = w_1 w_2 \cdots w_n$ be a freely reduced word in the generators $\{a, b, c\}$ and their inverses. In order to show that the even subgroup is free with basis $\{a, b, c\}$, we need to prove that $\omega \neq e \in \mathbb{F}_2$.

Notice that if you rewrite $\omega = w_1 w_2 \cdots w_n$ in terms of the original generators ($x$ and $y$) there may be cancellation. For example, $\omega$ may contain $a^{-1}b$, and expressing $a^{-1}b$ in terms of the original generators we get: $a^{-1}b = x^{-1}x^{-1}xy = x^{-1}y$. Similarly,

$$c^{-1}ac^{-1}b = (yx^{-1})(xx)(yx^{-1})(xy) = \left[yx^{-1}xx\right]\left[yx^{-1}xy\right] = yxy^2.$$

Notice that in this example, while there is cancellation in $c^{-1}a$ and $b^{-1}a$, there is no cancellation when one rewrites $ac^{-1}$ in terms of the original generators.

Converting $\omega$ from a freely reduced word in $\{a, b, c, a^{-1}, b^{-1}, c^{-1}\}^*$ to a word in $\{x, y, x^{-1}, y^{-1}\}^*$ is done by replacing each $w_i$ by $\alpha_i \beta_i$ where $\alpha_i, \beta_i \in \{x, y, x^{-1}, y^{-1}\}$. In the example above, with the word $c^{-1}ac^{-1}b$, we had $\beta_1 = x^{-1} = \alpha_2^{-1}$ and $\beta_3 = x^{-1} = \alpha_4^{-1}$. Because of this there was a small amount of internal cancellation, but it ends up that such constrained cancellations are the worst that can happen.

**Exercise 3.14.** Let $\omega = w_1 w_2 \cdots w_n$ be a freely reduced word in the free monoid $\{a, b, c, a^{-1}, b^{-1}, c^{-1}\}^*$ and let $\alpha_i \beta_i$ be the expression for $w_i$ using $\{x, y, x^{-1}, y^{-1}\}$. By considering cases, show that if $\beta_i = \alpha_{i+1}^{-1}$, then:

1. $\alpha_i \neq \beta_{i+1}^{-1}$;
2. $\beta_{i+1} \neq \alpha_{i+2}^{-1}$; and
3. $\beta_{i-1} \neq \alpha_i^{-1}$.

We can now show that the even subgroup is a free group with basis $\{a, b, c\}$. We first convert a given freely reduced word

$$\omega \in \{a, b, c, a^{-1}, b^{-1}, c^{-1}\}^*$$

to a word $\alpha_1 \beta_1 \cdot \alpha_2 \beta_2 \cdots \alpha_n \beta_n \in \{x, y, x^{-1}, y^{-1}\}^*$. Given 3.14, we then know that, when we reduce this word, at most one of the letters, $\alpha_i$ or $\beta_i$, is cancelled from each pair. Thus

$$|\alpha_1 \beta_1 \cdot \alpha_2 \beta_2 \cdots \alpha_n \beta_n| \geq n,$$

hence $\omega \neq e \in \mathbb{F}_2$.

### 3.3 Free Group Homomorphisms and Group Presentations

Given a group $G$ we describe how one can define a group homomorphism $\mathbb{F}_n \to G$ simply by picking targets for a basis of $\mathbb{F}_n$.

**Theorem 3.15.** *Let $G$ be any group and let $\{g_1, \ldots, g_n\}$ be a list of elements of $G$, which are not necessarily distinct or non-trivial. Let $S = \{x_1, \ldots, x_n\}$ be a basis for a free group $\mathbb{F}_n$. Then there is a group homomorphism $\phi : \mathbb{F}_n \to G$ where $\phi(x_i) = g_i$.*

*Proof.* We know where $\phi$ takes the basis $\{x_1, \ldots, x_n\}$ and if $\phi$ is to be a homomorphism it follows that $\phi(x_i^{-1})$ must be $g_i^{-1}$. If $\omega$ is an arbitary element of $\mathbb{F}_n$ then $\omega$ can be expressed as $\omega = w_1 w_2 \cdots w_k$ where each $w_i \in \{X \cup X^{-1}\}$. Define $\phi(\omega)$ to be

$$\phi(\omega) = \phi(w_1)\phi(w_2) \cdots \phi(w_n) \in G.$$

The elements of $\mathbb{F}_n$ are equivalence classes of words, where the equivalence relation is generated by

$$w_1 \cdots w_i x x^{-1} w_{i+1} \cdots w_k \sim w_1 \cdots w_i w_{i+1} \cdots w_k$$

where $x \in \{X \cup X^{-1}\}$. Since our definition of $\phi$ used a specific word to represent $\omega$, we need to make sure that the result would be the same if we had chosen a different word to represent $\omega$. However,

$$\phi(\omega) = \phi(w_1 w_2 \cdots w_k) = \phi(w_1)\phi(w_2) \cdots \phi(w_k)$$
$$= \phi(w_1)\phi(w_2) \cdots \phi(w_i)\phi(x)\phi(x)^{-1}\phi(w_{i+1}) \cdots \phi(w_k)$$
$$= \phi(w_1 w_2 \cdots w_i \cdot x \cdot x^{-1} \cdot w_{i+1} \cdots w_n).$$

Thus the element $\phi(\omega) \in G$ does not depend on which word is used to represent $\omega$. It follows immediately from the definition that $\phi(\omega^{-1}) = [\phi(\omega)]^{-1}$ and $\phi(\omega \cdot \omega') = \phi(\omega)\phi(\omega')$. So the function $\phi$ is indeed a group homomorphism. □

It is important to notice that the technique for constructing homomorphisms used in Theorem 3.15 does not work in general. For example, let $a$ and $b$ be two reflections that generate the infinite dihedral group $D_\infty$. There is no homomorphism $\phi : D_\infty \to \mathbb{Z}$ where $\phi(s) = \phi(t) = 1 \in \mathbb{Z}$. If there were, then $\phi(s^2) = 2$, but $s^2 = e \in D_\infty$, so $\phi$ would not take the identity element of $D_\infty$ to the identity element of $\mathbb{Z}$. The fact that this approach does work for free groups has a number of important consequences. One is the following corollary, which indicates why people refer to $\mathbb{F}_n$ as *the* free group of rank $n$ (and not as *a* free group of rank $n$).

**Corollary 3.16.** *Any two free groups of rank $n$ are isomorphic.*

*Proof.* Let $G$ be a free group with basis $\{x_1, \ldots, x_n\}$ and let $H$ be free with basis $\{z_1, \ldots, z_n\}$. By Theorem 3.15 there is a homomorphism $\phi : G \to H$ where $\phi(x_i) = z_i$ and a homomorphism $\psi : H \to G$ where $\psi(z_i) = x_i$. It follows that $\psi \circ \phi$ is the identity automorphism of $G$ and $\phi \circ \psi$ is the identity automorphism of $H$. Thus $\phi$ and $\psi$ must be bijections, and $G \approx H$. ☐

There is also the following result, which is the foundation for what are called group presentations.

**Corollary 3.17.** *If $G$ is generated by $n$ elements then $G$ is a quotient of $\mathbb{F}_n$.*

*Proof.* Theorem 3.15 establishes that there is a group homomorphism $\phi : \mathbb{F}_n \to G$ that takes a basis of $\mathbb{F}_n$ to a generating set for $G$. Since the image of $\phi$ contains the generators of $G$, the image must be all of $G$. Thus, by the First Isomorphism Theorem, $G \approx \mathbb{F}_n / \ker(\phi)$. ☐

Group presentations are not something that we will explore in earnest in this book but, as we are so close to the basic definition, we might as well take a minor detour and introduce the topic.

Let $G$ be a finitely generated group, with a chosen set of generators $S = \{g_1, \ldots, g_n\}$. Then by Corollary 3.17 there is a homomorphism $\phi$ from the free group $\mathbb{F}_n$ with basis $X = \{x_1, \ldots, x_n\}$ onto $G$, induced by requiring $\phi(x_i) = g_i$. If $\omega$ is a word that maps to the identity in $G$ under $\phi$, then $\omega$ is called a *relation*. This is a natural phrase, as such words give rise to equations of the form

"some product of generators and their inverses" $= e \in G$.

For example, if $a$ and $b$ generate $\mathbb{Z} \oplus \mathbb{Z}$ then $aba^{-1}b^{-1}$ is a relation.

**Remark 3.18.** Because relations are words that represent the identity, there is a one-to-one correspondence between relations and paths in the Cayley graph that begin and end at the vertex associated to the identity.

Let $G$ be a group and $\phi : \mathbb{F}_n \to G$ a surjective map as above. A subset $R \subset \ker(\phi)$ is a *set of defining relations* if the smallest normal subgroup of $\mathbb{F}_n$ that contains $R$ is $\ker(\phi)$. Note that the smallest normal subgroup containing $R$ must contain both $R$ and $R^{-1}$ (because it is a subgroup) as well as every conjugate $grg^{-1}$ for $g \in \mathbb{F}_n$ and $r \in R \cup R^{-1}$ (because it is normal). In fact, the set of all finite products of all conjugates of the elements of $R \cup R^{-1}$ is the smallest normal subgroup containing $R$. Thus if $R$ is a set of defining relations then every element in the kernel can be expressed as a finite product of conjugates of the elements of $R$ and their inverses.

A finitely generated group $G$ is said to be *finitely presented* if there is a finite set of defining relations, $R$. In this case one indicates these generators and defining relations by

$$G = \langle g_1, \ldots, g_n \mid \omega_1, \ldots, \omega_m \rangle$$

where $\{g_1, \ldots, g_n\}$ is the generating set and $R = \{\omega_1, \ldots, \omega_m\}$. In general, every group has a presentation like that given above, except that the list of generators and/or the list of relations may not be finite. On the other hand, most finitely generated groups that you have encountered to date are finitely presented. For example, $\mathbb{Z} \oplus \mathbb{Z} = \langle a, b \mid aba^{-1}b^{-1} \rangle$, which is often written $\langle a, b \mid ab = ba \rangle$. A standard presentation of the infinite dihedral group is

$$D_\infty = \langle a, b \mid a^2, b^2 \rangle$$

and the finite dihedral groups have presentations

$$D_n = \langle a, b \mid a^2, b^2, (ab)^n \rangle.$$

We are certainly not claiming these presentations are obvious; it takes some effort to establish that a given set of relations forms a set of defining relations. A good introduction to the topic of group presentations is Johnson's *Presentations of Groups* [Jo97].

Group presentations are not just of interest in the study of infinite groups. They are used, for example, in Conway and Lagarias's proof of the following combinatorial theorem:

**Theorem 3.19** (See [CL90] and [Th90]). *Let $T_n$ be a triangle of hexagons, with $n$ hexagons on each side, as in Figure 3.7. Then*

1. *$T_n$ can be tiled by $T_3$ if and only if*

$$n \equiv 0, 2, 9 \ or \ 11(mod \ 12);$$

2. *$T_n$ cannot be tiled by the line of hexagons $L_3$ for any $n$.*

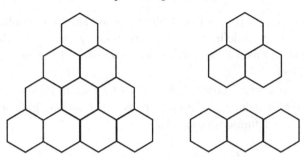

Fig. 3.7. Two triangles of hexagons are shown: On the left is $T_4$ and on the top right is $T_2$. On the bottom right is the line of hexagons $L_3$.

## 3.4 Free Groups and Actions on Trees

The following result characterizes free groups via actions on trees. It gives insight into the nature of free groups and it provides a useful approach for proving results, notably the Nielsen–Schreier Theorem (Corollary 3.23).

**Theorem 3.20.** *A group $G$ is free if and only if it acts freely on a tree.*

In Section 1.8 we proved that if a group acts on a graph then there is a fundamental domain for that action (see Lemma 1.52 in particular). We briefly recall the construction of this fundamental domain in the context of a group $G$ acting freely on a tree $T$. Start with a fixed vertex $v \in T$, and consider subtrees $C \subset T$ where

1. $v \in C$; and
2. if $x$ and $y$ are distinct vertices in $C$ then there is no element $g \in G$ such that $g \cdot x = y$.

There is at least one maximal subtree satisfying these conditions, and we call it CORE. Because the action of $G$ is free, the subtree CORE does not form a fundamental domain; the image of CORE under the action of $G$ contains every vertex of $T$, but it does not contain every edge. For each edge $e \not\subset$ CORE with $e \cap$ CORE $\neq \emptyset$, let $h_e$ be the closed half of $e$ that is connected to CORE. Define $\mathcal{F}$ to be the union of CORE and these half-edges. The proof of Lemma 1.52 shows:

**Lemma 3.21.** *The subset $\mathcal{F} \subset T$ described above is a fundamental domain for the action $G \curvearrowright T$, and $\mathcal{F}$ contains one vertex for each orbit of vertices under the action of $G$.*

*Proof of Theorem 3.20.* One direction is easy: if $G$ is a free group, then it acts freely on its Cayley graph, which is a tree.

Assume then that $G$ acts freely on a tree $T$ and that $\mathcal{F}$ is a fundamental domain as described above. For each half-edge $h_e \in \mathcal{F}$ let $g_e \in G$ be the element such that $\mathcal{F} \cap g_e \cdot \mathcal{F}$ is the midpoint of $e$. Denote the set of all such $g_e$ by $S$. Since $\mathcal{F} \cap g_e \cdot \mathcal{F}$ is the midpoint of an edge, $g_e^{-1} \cdot \mathcal{F} \cap \mathcal{F}$ is also the midpoint of an edge. Thus if $g_e \in S$ so is $g_e^{-1}$, hence:

<u>Claim:</u> *If $g_e \in S$ then there is a half-edge $h_{\hat{e}} \in \mathcal{F}$ such that $g_e^{-1} = g_{\hat{e}}$.*

The set of elements $S$ is a set of generators of $G$ by Theorem 1.55. As each element in $S$ can be paired with its inverse, we may take $S$ to be a maximal subset of $S$ which does not contain both an element and its inverse. Thus $S = S \cup S^{-1}$.[3]

Fix a vertex $v \in$ Core and for each $g_e \in S \cup S^{-1}$ let $T_e$ be the subtree of $T$ induced by the set of vertices $v' \in T \setminus$ Core such that the reduced path from $v$ to $v'$ passes through the half-edge $h_e$. Equivalently one could define $T_e$ to be the maximal subtree contained in $T \setminus$ Core that shares a vertex with the edge $e$. For example, take the action of $\mathbb{F}_2$ on its Cayley graph. Then Core is a single vertex and the fundamental domain $\mathcal{F}$ consists of this vertex plus four half-edges. The set $S$ will consist of two elements, and the four subtrees $T_e$ are the maximal subtrees contained in the connected components formed by removing $\mathcal{F}$ (see Figure 3.8).

We may now apply the Ping Pong Lemma (Lemma 3.10) to prove that $G$ is a free group with basis $S$. Using the notation of the Ping Pong Lemma:

- for each $g_e \in S$, let $X_s$ be the associated subtree $T_e$;
- take $p = v \in$ Core to be a point outside of the union of the subtrees $\bigcup T_e$.

Condition 1 of the Ping Pong Lemma follows from the definition of the subtrees $T_e$. Let $g_e \in S$ and let $T_{e'}$ be one of the subtrees, where $g_{e'} \neq g_e^{-1}$. Let $\hat{e} = g_e^{-1} \cdot e$ denote the edge connecting $g_e^{-1}($Core$)$ to Core. Given any vertex $w \in T_{e'}$ there is a minimal edge path connecting $g_e^{-1} \cdot v$ to $w$. Applying $g_e$ we get a minimal edge path from $v$ to $g_e \cdot w$. The original path passed through $\hat{e} = g_e^{-1} \cdot e$, hence the the image under

---

[3] We have cheated in this paragraph. We have assumed that there is no $g \in G$ such that $g = g^{-1}$. In other words, we are assuming there is no element of order 2 in $G$. However, Theorem 3.46 states that when a finite group acts on a tree, it must fix some point in that tree. It follows that if $G$ contains an element of order 2, then $G$ cannot act freely on a tree. The reader worried about circularity should work through the proof of 3.46 before proceeding.

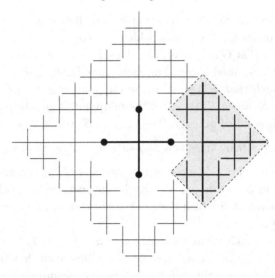

Fig. 3.8. A fundamental domain for the action of $\mathbb{F}_2$ on its Cayley graph is a cross of four half-edges. For each of these half-edges, there is a corresponding subset $\mathcal{T}_e \subset \mathcal{T}$. One such subtree is highlighted above.

$g_e$ will pass through $e$. Thus $g_e \cdot w \in \mathcal{T}_e$ by construction. Since $w$ was an arbitrary vertex of $\mathcal{T}_{e'}$, $g_e(\mathcal{T}_{e'}) \subset \mathcal{T}_e$, establishing Condition 2.  $\square$

A careful reading of the proof of Theorem 3.20 shows that it is constructive; it provides a method for finding a basis for $G$.

**Example 3.22.** Consider the even subgroup $H < \mathbb{F}_2$, where $\{x, y\}$ is a basis of $\mathbb{F}_2$, discussed in Proposition 3.12. The elements of $H$ consist of all elements of $\mathbb{F}_2$ of even length. This subgroup acts on the Cayley graph $\Gamma$ of $\mathbb{F}_2$ and we may begin constructing a core by selecting the vertex associated to the identity to be part of CORE. As we also need a vertex representing elements of odd length, we add the vertex associated to $x$. The edge $e$ joining these two vertices completes the construction of CORE. The fundamental domain $\mathcal{F}$ consists of this edge $e$ and the six half-edges incident with it.

Since there are six half-edges, we immediately know that our basis will have three elements. To find a basis, we need to find all the $g \in H$ where $g \cdot \mathcal{F} \cap \mathcal{F} \neq \emptyset$. The complete list of such elements is

$$\{x^2, x^{-2}, xy, y^{-1}x^{-1}, xy^{-1}, yx^{-1}\}.$$

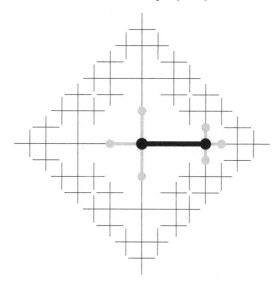

Fig. 3.9. A Core and fundamental domain for the action of the even subgroup on the Cayley graph of $\mathbb{F}_2$.

After removing inverses we see that $\{x^2, xy, xy^{-1}\}$ is a basis for the even subgroup of $\mathbb{F}_2$.

We end this section with a striking result about free groups.

**Corollary 3.23** (Nielsen–Schreier Theorem). *Every subgroup of a free group is free.*

*Proof.* Let $H$ be a subgroup of a free group $\mathbb{F}$. The group $\mathbb{F}$ acts freely on its Cayley graph, and therefore $H$ acts freely on the Cayley graph of $\mathbb{F}$. Thus, by the characterization of free groups given by Theorem 3.20, $H$ is a free group. $\qquad\square$

This result was first proved by Jakob Nielsen for finitely generated free groups in 1921. His result was extended to all free groups by Otto Schreier a few years later.

## 3.5 The Group $\mathbb{Z}_3 * \mathbb{Z}_4$

The infinite dihedral group, $D_\infty$, can be thought of as a subgroup of $\mathrm{SYM}^+(\mathcal{T}_{2,2})$, where $\mathcal{T}_{2,2}$ is the bipartite tree where every vertex has valence 2. (In fact, in Exercise 2 of Chapter 2 you were asked to prove

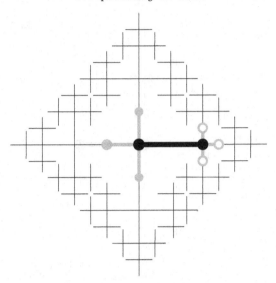

Fig. 3.10. The element $x^2$ can be chosen to be part of a basis for the even subgroup of $\mathbb{F}_2$ since $x^2 \cdot \mathcal{F}$ intersects $\mathcal{F}$ at the rightmost point of $\mathcal{F}$. Because $xy \cdot \mathcal{F}$ intersects $\mathcal{F}$ at the highest point highlighted above, and $xy^{-1} \cdot \mathcal{F}$ intersect $\mathcal{F}$ at the lowest point highlighted above, we know that $\{x^2, xy, xy^{-1}\}$ is a basis.

that $D_\infty \approx \mathrm{SYM}^+(\mathcal{T}_{2,2})$.) In this section we explore a similar group, defined in terms of symmetries of the biregular tree $\mathcal{T}_{3,4}$. Let $a$ be a rotation about a vertex of valence 4 in $\mathcal{T}_{3,4}$ and let $b$ be a rotation about an adjacent vertex of valence 3. Let $G$ be the subgroup of $\mathrm{SYM}(\mathcal{T}_{3,4})$ generated by $a$ and $b$.

In order to avoid confusion about what is meant by a "rotation" of $\mathcal{T}_{3,4}$, we introduce labels for the vertices. Every vertex in $\mathcal{T}_{3,4}$ can be reached by a unique, reduced path starting at the vertex fixed by $a$. Such a path starts by going <u>N</u>orth, <u>E</u>ast, <u>S</u>outh or <u>W</u>est. Then one chooses to go either <u>L</u>eft or <u>R</u>ight. After the <u>L</u>eft/<u>R</u>ight choice, there are three directions available that do not involve backtracking. If we always pretend that the edge we are on corresponds to <u>E</u>ast, keep the clockwise orientation and use the pneumonic "<u>N</u>ever <u>E</u>at <u>S</u>our <u>W</u>atermelons," then we can unambiguously move forward by going <u>S</u>outh, <u>W</u>est or <u>N</u>orth. Every vertex of $\mathcal{T}_{3,4}$ then has a unique label involving an alternating sequence of cardinal directions and left/right choices. These coordinates are illustrated in Figure 3.12. The only vertex that does not have a label is the vertex fixed by $a$, which we will refer to as "the origin."

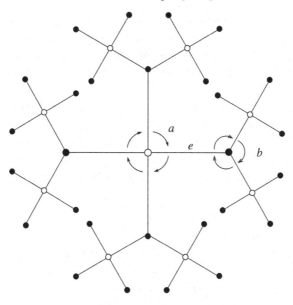

Fig. 3.11. The "rotations" $a$ and $b$ in SYM($\mathcal{T}_{3,4}$).

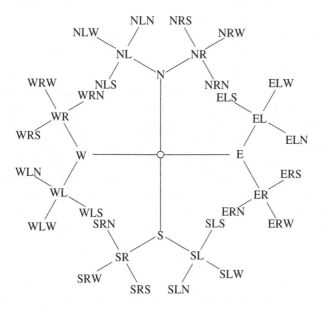

Fig. 3.12. Labelling the vertices of $\mathcal{T}_{3,4}$ using left/right and the four cardinal directions.

The element $a$ fixes the origin, and otherwise it applies the permutation $N \Rightarrow E \Rightarrow S \Rightarrow W \Rightarrow N$ to the first letter labelling a vertex $v$. So, as examples, under the action of $a$ the vertex labelled NRW is taken to the vertex labelled ERW and the vertex labelled WLS is taken to the vertex labelled NLS.

The action of $b$ is a bit more complicated to describe in terms of these coordinates. In order to avoid repeatedly writing "the vertex labelled by __" we conflate vertices with their labels. The vertex E is fixed by $b$. The origin is taken to EL; the vertex EL is taken to ER; and ER is taken to the origin. The rules for determining the action of $b$ on the remaining vertices are:

1. If a vertex has "EL" as a prefix, it is taken to the vertex obtained by replacing this prefix by "ER."
2. If a vertex has "ER" as a prefix, strip off this prefix.
3. If $v$ has any other prefix, then add EL as a prefix. (The vertex WL is taken to ELWL, for example.)

To get a little bit of practice with this group action, you should show:

1. The element $aba^{-1}$ is a rotation about the vertex labelled S in Figure 3.12.
2. The element $a^2ba^{-2}$ is a rotation about the vertex W.
3. The element $a^3ba^{-3}$ is a rotation about the vertex N.

The group $G$, generated by the rotations $a$ and $b$, is referred to as the *free product* of $\mathbb{Z}_3$ and $\mathbb{Z}_4$, and it is denoted $\mathbb{Z}_3 * \mathbb{Z}_4$. The $\mathbb{Z}_3$ refers to the fact that $b \in G$ generates a cyclic subgroup of order 3 and the $\mathbb{Z}_4$ refers to the fact that $a \in G$ generates a cyclic subgroup of order 4. (We discuss free products of groups in general in the next section.)

Because $a$ and $b$ generate $\mathbb{Z}_3 * \mathbb{Z}_4$, and $a$ has order 4 and $b$ has order 3, any $g \in \mathbb{Z}_3 * \mathbb{Z}_4$ can be expressed as a product $g = a^{i_1}b^{j_1}\cdots a^{i_n}b^{j_n}$ where $1 \le i_k \le 3$ and $1 \le j_k \le 2$, except possibly $i_1 = 0$ and/or $j_n = 0$. Define the *norm*[4] of such an expression to be the number of non-zero exponents, and denote this by $||s^{i_1}t^{j_1}\cdots s^{i_n}t^{j_n}||$. For example,

$$||st^3s^2tst^3s^2|| = 7 .$$

Let $e$ be the edge joining the vertex fixed by $a$ to the vertex fixed by $b$ or, equivalently, the edge joining the origin and the vertex labelled E

---

[4] What we are calling the "norm" is also frequently referred to as the "syllable length" of a word.

in Figure 3.12. By reading a word $\omega = a^{i_1} b^{j_1} \cdots a^{i_n} b^{j_n}$ left-to-right one can easily determine the image of $e$ under the action of $\omega$.

Assuming that $i_1 \neq 0$, one rotates $e$ about the white vertex fixed by $a$ as far as $i_1$ dictates. Denote the resulting edge $e'$. Rotate $e'$ around its black vertex as far as is dictated by $j_1$. Denote the resulting edge $e''$, and then rotate $e''$ about its white vertex as far as is dictated by $i_2$, and so on. Similarly, if $i_1 = 0$, and so $\omega$ really should be expressed as $\omega = b^{j_1} a^{i_2} \cdots a^{i_n} b^{j_n}$, then one applies the same process, starting by rotating about the black vertex of $e$ as far as $j_1$ dictates, then rotating about the resulting white vertex as far as $i_2$ dictates, and so on. For example, if $\omega = a^2 ba$ then the image of $e$ under the action of $\omega$ is the edge $f$ shown in Figure 3.13.

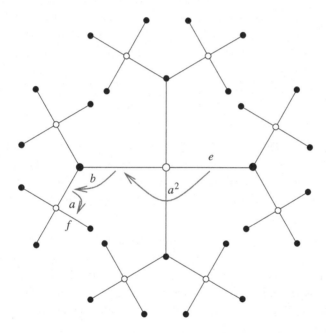

Fig. 3.13. The element $a^2 ba$ in $\mathbb{Z}_3 * \mathbb{Z}_4$ takes the edge $e$ to the edge $f$. The arrows are labelled using the "two wrongs make a right" principle.

You should be very suspicious of the previous description. Mixing up right and left can lead to errors when one is dealing with non-abelian groups. Yet it appears that we have done just that and we have also seemingly confused our definitions. In considering $\omega = a^2 ba$ it appears that we are letting "$b$" denote a rotation about the vertex labelled W.

But $b$ actually rotates about the vertex labelled E. These two errors together give the right answer by the "two wrongs make a right" principle that we previously encountered in our discussion of reflection groups (Chapter 2). You can see why this works if you notice that $\omega$ can be rewritten as a product of conjugates:

$$\omega = a^2ba = (a^2bab^{-1}a^{-2})(a^2ba^{-2})a^2$$

The rightmost $a^2$ gives the rotation about the unlabelled vertex; the term $(a^2ba^{-2})$ is the rotation falsely labelled "$b$" in Figure 3.13; the final expression $(a^2bab^{-1}a^{-2})$ is the rotation falsely labelled as "$a$" Figure 3.13.

This "two wrongs make a right" principle is actually just an application of a broader principle, namely, "conjugation is doing the same thing somewhere else." If $g \in \mathbb{Z}_3 * \mathbb{Z}_4$ takes the unlabelled vertex fixed by $a$ to some vertex $v$, then $ga^jg^{-1}$ will be a rotation about $v$. Thus $ga^j = (ga^jg^{-1})g$ will take the unlabelled vertex to $v$ (via $g$) and then rotate about $v$ (via $ga^jg^{-1}$).

This description of how to determine where the edge $e$ goes under the action of $\omega = a^{i_1}b^{j_1}\cdots a^{i_n}b^{j_n}$ is the key insight for establishing the following lemma.

**Lemma 3.24.** *The group $\mathbb{Z}_3 * \mathbb{Z}_4$ acts transitively on the edges of $\mathcal{T}_{3,4}$. Further, let $e$ be the edge joining the vertex stablized by $a$ to the vertex stabilized by $b$, and let $g = a^{i_1}b^{j_1}\cdots a^{i_n}b^{j_n}$ (as prescribed above) with $n = \|a^{i_1}b^{j_1}\cdots a^{i_k}b^{j_k}\|$. Then the number of vertices between $e$ and $g \cdot e$ is $n$.*

*Proof.* Given any edge $f$ we can use the method described above to write down an element of $g_f \in \mathbb{Z}_3 * \mathbb{Z}_4$ that will take $e$ to $f$. Thus the product $g_f \cdot g_{f'}^{-1}$ will take an edge $f'$ to $f$, so the action is edge-transitive. The claim about the norm determining the distance is immediate from the description of how one finds the image of $e$ under the action of $g$. $\square$

**Proposition 3.25.** *The action of $\mathbb{Z}_3 * \mathbb{Z}_4$ on $\mathcal{T}_{3,4}$ is simply transitive on the edges. Further, any $g \in \mathbb{Z}_3 * \mathbb{Z}_4$ can be uniquely expressed as a word $\sigma^{s_1}\tau^{t_1}\cdots\sigma^{s_n}\tau^{t_n}$ where $s_1$ and/or $t_n$ can be zero, but otherwise $1 \leq s_i \leq 2$ and $1 \leq t_i \leq 3$.*

*Proof.* That the action is transitive is established by Lemma 3.24. To establish that it is simply transitive, it suffices to show that no element of $\mathbb{Z}_3 * \mathbb{Z}_4$ can have two such expressions:

$$g = s^{i_1}t^{j_1}\cdots s^{i_n}t^{j_n} = s^{k_1}t^{l_1}\cdots s^{k_m}t^{l_m}.$$

If there was such a pair of expressions, then

$$s^{i_1}t^{j_1}\cdots s^{i_n}t^{j_n}\cdot t^{-l_m}s^{-k_m}\cdots t^{-l_1}s^{-k_1}=e.$$

But then by Lemma 3.24 the norm of this product must be zero. This is only possible if $t^{j_n}\cdot t^{-l_m}=e$, that is, $j_n=l_m$, and then $s^{i_n}\cdot s^{-k_m}=e$, hence $i_n=k_m$, and so on until there has been complete cancellation. This contradicts the assumption that $g$ has two distinct expressions. $\quad\Box$

## 3.6 Free Products of Groups

A group $G$ is the direct product of two subgroups, $A$ and $B$, if both $A$ and $B$ are normal, $G=A\cdot B$, and $A\cap B=e$. It makes sense to refer to $G$ as a product, since in essence it is produced directly from the groups $A$ and $B$. A distinguishing feature of the direct product is the fact that elements of $A$ commute with elements of $B$ in $G=A\oplus B$.

In this section we introduce the free product of two groups $A$ and $B$, denoted $A*B$. The free product of $A$ and $B$ can be thought of as a group generated by a copy of $A$ and a copy of $B$ where there is no interaction – such as commuting – between the subgroups $A$ and $B$.

To describe the elements of the free product $A*B$ we start with an alphabet that is the union of the elements of $A$ and $B$, $\{A\cup B\}$. A word $\omega=x_1x_2\cdots x_n\in\{A\cup B\}^*$ is a *reduced word* if

1. no $x_i=e_A$ or $e_B$ (where $e_A$ is the identity element of $A$ and $e_B$ is the identity element of $B$); and
2. when $x_i\in A$ then $x_{i+1}\notin A$, and similarly if $x_i\in B$ then $x_{i+1}\notin B$.

In other words, a reduced word is one that alternates between non-identity elements of $A$ and $B$.

We would like to take the set of reduced words to be the elements of the free product $A*B$, with the binary operation being concatenation. But there is a minor difficulty, in that the concatenation of two reduced words might not be reduced. Thus we introduce the following equivalence relation. Let $\sim$ be the equivalence relation on $\{A\cup B\}^*$ generated by

1. $\omega e_x\omega'\sim\omega\omega'$ where $e_x$ is either $e_A$ or $e_B$;
2. $\omega\cdot ab\cdot\omega'\sim\omega\cdot c\cdot\omega'$ where $ab=c$ holds in either the group $A$ or the group $B$.

We define the elements of $A * B$ to be the equivalence classes $[\omega]$ of words $\omega \in \{A \cup B\}^*$. The binary operation is defined by taking the equivalence class of the concatenation of two words: $[\omega_1][\omega_2] = [\omega_1\omega_2]$. (The reader should check that this definition does not depend on the choice of representatives of the equivalence classes.) The identity element is the empty string, and if $\omega = x_1 x_2 \cdots x_n$ then the inverse of $[\omega]$ is the equivalence class of its formal inverse $\omega^{-1} = x_n^{-1} \cdots x_1^{-1}$.

**Theorem 3.26.** *Each equivalence class $[\omega]$ for $\omega \in \{A \cup B\}^*$ contains exactly one reduced word.*

*Proof.* We first prove that every word $\omega$ is equivalent to a reduced word $\varrho(\omega)$. We do this by induction on the length of the word $\omega$. If $\omega$ is the empty word or is of length one, then $\omega$ is reduced and we may take $\varrho(\omega) = \omega$. For any $x \in A \cup B$ and any reduced word $\omega = x_1 \cdots x_n$ define the reduced word $\varrho(x\omega)$ by

$$
\varrho(x\omega) = \begin{cases}
\omega & \text{if } x = e_A \text{ or } e_B \\
x\omega & \text{if } x \in A \setminus \{e_A\} \text{ and } x_1 \in B \\
& \text{or } x \in B \setminus \{e_B\} \text{ and } x_1 \in A \\
yx_2 \cdots x_n & \text{if } x, x_1 \in A \text{ and } x \cdot x_1 = y \neq e_A \\
& \text{or } x, x_1 \in B \text{ and } x \cdot x_1 = y \neq e_B \\
x_2 \cdots x_n & \text{if } x = x_1^{-1}.
\end{cases}
$$

For an arbitrary word $\omega$ we may now define $\varrho(\omega)$ inductively. Write $\omega = x\omega'$, where $x$ is the leading letter of $\omega$, and set $\varrho(\omega) = \varrho(x\varrho(\omega'))$. By induction $\omega' \sim \varrho(\omega')$ and therefore $x\omega' \sim x\varrho(\omega')$. Thus we have $\omega = x\omega' \sim \varrho(x\varrho(\omega'))$, and this last expression is by definition a reduced word.

The argument that there is a unique reduced word in each equivalence class mimics the proof of Cayley's Basic Theorem. Let $\Omega_R$ be the collection of all reduced words in $\{A \cup B\}^*$. Let $x \in A \cup B$ and let $\pi_x$ be the permutation of $\Omega_R$ described by $\pi_x : \omega \mapsto \varrho(x\omega)$. For any word $\omega = x_1 \cdots x_n$ define the permutation $\pi_\omega$ as $\pi_\omega = \pi_{x_1} \cdots \pi_{x_n}$. By considering the generating conditions (1) and (2) above, one can show that if $\omega \sim \omega'$ then $\pi_\omega$ is the same permutation as $\pi_{\omega'}$. Since the elements of $A * B$ are the equivalence classes of words, it follows that we have constructed a representation $\pi : A * B \to \text{SYM}(\Omega_R)$.

To see that no equivalence class contains two distinct reduced words, it suffices to show that two distinct reduced words induce distinct permutations. (In other words, we show that the representation above is faithful.) But if $\omega$ is any reduced word, and $\emptyset$ represents the empty word, then $\pi_\omega(\emptyset) = \omega$. Thus if $\omega_1$ and $\omega_2$ are distinct reduced words, then $\pi_{\omega_1} \neq \pi_{\omega_2}$, as these permutations do not agree on the reduced word $\emptyset$. So no two distinct, reduced words can represent the same element of $A * B$. $\qquad\square$

**Corollary 3.27.** *The groups $A$ and $B$ inject into $A * B$.*

*Proof.* The elements of $A$ and $B$ form reduced words (of length 1). $\quad\square$

**Theorem 3.28.** *Every free product of groups $A * B$ can be realized as a group of symmetries of a biregular tree $\mathcal{T}$, and a fundamental domain for this action consists of a single edge and its two vertices. Further, if $A$ and $B$ are both finite, this tree is $\mathcal{T}_{|A|,|B|}$.*

*Proof.* Define $V(\mathcal{T})$ to be the the left cosets $gA$ and $gB$. In order to have a bit of notation, we let $v_{gA}$ denote the vertex associated to the coset $gA$ and similarly $v_{gB}$ is the vertex associated to $gB$.

The edges of the tree $\mathcal{T}$ correspond to the elements of $A * B$, and we use $e_g$ to denote the edge associated to $g \in A * B$. This does have the unfortunate consequence that the edge associated to the identity is "$e_e$." We define $\mathrm{ENDS}(e_g) = \{v_{gA}, v_{gB}\}$. The bipartite structure of $\mathcal{T}$ comes from considering the vertices associated to cosets of $A$ as one "part" and the vertices associated to cosets of $B$ as the other "part."

Note that two edges $e_g$ and $e_h$ intersect if and only if $g^{-1}h \in A$ or $B$. For example, if $e_g \cap e_h = v_{kA}$ then $g \in kA$ and $h \in kA$, hence $g^{-1}h \in A$. Conversely, if $g^{-1}h \in A$ then $gA = hA$ and so $v_{gA} = v_{hA}$. The same argument holds *mutatis mutandis* for $B$.

The action of $A * B$ on $\mathcal{T}$ is induced by the standard left action of a group on itself and on its left cosets. An element $g \in A * B$ moves vertices by $g : v_{hA} \mapsto v_{ghA}$ and $g : v_{hB} \mapsto v_{ghB}$; the edges move by $g : e_h \mapsto e_{gh}$. As in the case of the action of a group on any of its Cayley graphs, we get an action $G \curvearrowright \mathcal{T}$ primarily because the group is permuting labels by left multiplication, while incidence is defined by right multiplication. It follows from the description of this action that the edge $e_e$, which joins $v_A$ to the $v_B$, is a fundamental domain for the action of $A * B$ on $\mathcal{T}$.

To prove that the graph $\mathcal{T}$ is a tree we need to show that it is connected and contains no cycles. If $g \in A * B$ then $g = [\omega]$ for a reduced word $\omega = x_1 x_2 \cdots x_n$. Assume for convenience that $x_1 \in A$, $x_2 \in B$, and so on. Then $e_e \cap e_{x_1} = v_A$, $e_{x_1} \cap e_{x_1 x_2} = v_{x_1 B}$, and so on. Hence $\{e_e, e_{x_1}, e_{x_1 x_2}, \ldots, e_{x_1 x_2 \cdots x_n}\}$ is an edge path from $e_e$ to $e_g$. It follows that any edge $e_g$ can be joined to $e_e$ by an edge path, hence $\mathcal{T}$ is connected.

We use Theorem 3.26 to show that $\mathcal{T}$ has no circuits. If there is a circuit in $\mathcal{T}$, then using the action $A * B \curvearrowright \mathcal{T}$ we may move this to a circuit that begins with the edge corresponding to the identity: $\{e_e, e_1, \ldots, e_n, e_e\}$. Then $e_1 = e_{x_1}$ for some $x_1 \in A$ or $B$ and, without loss of generality, we will assume $x_1 \in A$. Then $e_2 = e_{x_1 x_2}$, for some $x_2 \in B$, and in general $e_k = e_{x_1 x_2 \cdots x_k}$ where $x_1 x_2 \cdots x_k$ is a reduced word in $\{A \cup B\}^*$. Since this edge path is assumed to be a circuit, $e_{x_1 x_2 \cdots x_n} \cap e_e = v_A$ or $v_B$. Hence $x_1 x_2 \cdots x_n = x$ for $x \in A \cup B$. But this is impossible, as this would imply that we have two distinct reduced words in an equivalence class.

To establish the final claim, notice that $v_{gA} \in \text{ENDS}(e_h)$ if and only if $h \in gA$. Thus the number of edges incident with $v_{gA}$ is the size of the coset $gA$, which is the same as the order of $A$. The same argument shows that the degree of vertices of the form $v_g B$ is $|B|$.  □

Let's revisit the example of $\mathbb{Z}_3 * \mathbb{Z}_4$ considered in the previous section. The edges of the tree $\mathcal{T}$ associated to a free product $A*B$ (Theorem 3.28) correspond to elements of $A * B$, which in turn correspond to reduced words. Our labelling of the vertices of $\mathcal{T}_{3,4}$ in Section 3.5 secretly exploited this fact. One can see this by making the substitutions

$$\text{E} \mapsto e, \quad \text{S} \mapsto a, \quad \text{W} \mapsto a^2, \text{ and } \text{N} \mapsto a^3$$

along with

$$\text{L} \mapsto b, \text{ and } \text{R} \mapsto b^2.$$

The edge joining the origin to the vertex associated to $E$ is the edge $e_e$ from Theorem 3.28; the edge joining the origin to $S$ is $e_a$; and the edge joining $WL$ to $WLS$ is $e_{a^2 ba}$, as is shown in Figure 3.13. The action of $a$ and $b$ on $\mathcal{T}_{3,4}$, described in Section 3.5 in terms of adjusting labels, can be reconstructed by simply applying the above translation scheme with left multiplication.

**Exercise 3.29.** Pick two finite groups, $A$ and $B$, that you are comfortable with. Construct the tree $\mathcal{T}$, and the action $A * B \curvearrowright \mathcal{T}$, given by Theorem 3.28. Be as concrete as possible.

**Corollary 3.30.** *The stabilizers of the vertices of the tree associated to a free product $A * B$ are conjugates of $A$ and $B$.*

*Proof.* Let $h$ be an element of the subgroup $gAg^{-1}$. Then

$$h \cdot v_{gA} = gag^{-1} \cdot v_{gA} = ga \cdot v_A = g \cdot v_A = v_{gA}.$$

Thus $gAg^{-1}$ is a subgroup of the stabilizer of $v_{gA}$, $\mathrm{Stab}(v_{gA})$. A similar computation shows that $g^{-1}\mathrm{Stab}(v_{gA})g < A$. Thus

$$A < g^{-1}\mathrm{Stab}(v_{gA})g < A$$

hence $\mathrm{Stab}(v_{gA}) = gAg^{-1}$.

The same argument shows that $\mathrm{Stab}(v_{gB}) = gBg^{-1}$. $\qquad\qquad\square$

**Exercise 3.31.** Let $e$ be any edge in the tree $\mathcal{T}$ associated to a free product $A * B$. Show that $\mathrm{Stab}(e)$ is trivial.

**Remark 3.32.** Theorem 3.26, which establishes a one-to-one correspondence between elements of $A * B$ and reduced words, comes from work of van der Waerden in the late 1940s. The idea of free products, however, dates at least to the 1920s, and was expressed quite clearly in a foundational paper by Kurosch in 1933 [Ku33]:

Die Gruppe $G$ heißt *freies Produkt* ihrer Untergruppen $H_\alpha$ ... wenn jedes Element $g$ von $G$ auf eine und nur eine Weise als Product

$$g = h_1 h_2 \cdots h_n, \ h_i \neq 1, h_i \in H_{\alpha_i}, \ H_{\alpha_i} \neq H_{\alpha_{i+1}}$$

darstellbar ist; es gibt also in $G$ keine nichtidentische Relationen, welche Elemente von verschiedenen $H_\alpha$ verbinden.

## 3.7 Free Products of Finite Groups are Virtually Free

If $\mathcal{P}$ is a group-theoretic property, then a group $G$ is said to be *virtually $\mathcal{P}$* if it has a finite-index subgroup $H < G$ where $H$ is a $\mathcal{P}$-group. For instance, a group $G$ is *virtually abelian* if it contains an abelian subgroup of finite-index. Example 3.36 shows there is a finite-index subgroup of $D_\infty$ that is isomorphic to $\mathbb{Z}$, hence the infinite dihedral group is *virtually cyclic*. The goal of this section is to prove that every free product of finite groups is *virtually free*, or, in other words, that such free products contain finite-index subgroups which are free.

Corollary 3.27 says that $A$ and $B$ inject into the free product $A * B$. Let $\iota_A : A \to A * B$ be the homomorphism sending $a \in A$ to the element of $A * B$ corresponding to the reduced word "$a$"; let $\iota_B : B \to A * B$ be the same sort of map for $B$. Exercise 19 asks you to provide a proof of the following theorem, which is the analog of Theorem 3.15 in the context of free products.

**Theorem 3.33.** *If there are group homomorphisms $\phi_A : A \to G$ and $\phi_B : B \to G$ then there is a map $\phi : A * B \to G$ such that $\phi_A = \phi \circ \iota_A$ and $\phi_B = \phi \circ \iota_B$.*

**Example 3.34.** Let $x$ generate a group $A$ of order 2 and let $y$ generate another group $B$ which is also of order 2. Then $x$ and $y$ (or more accurately $\iota_A(x)$ and $\iota_B(y)$) generate $\mathbb{Z}_2 * \mathbb{Z}_2$. Denote the generators of the infinite dihedral group, $D_\infty$, as $a$ and $b$, just as in Chapter 2. By Theorem 3.33 there is a homomorphism $\phi : \mathbb{Z}_2 * \mathbb{Z}_2 \to D_\infty$ induced by declaring $\phi(x) = a$ and $\phi(y) = b$. Since $a$ and $b$ generate $D_\infty$ this homomorphism is surjective. The map $\phi$ is injective because every $g \in \mathbb{Z}_2 * \mathbb{Z}_2$ can be represented by a reduced word, which in this case means a word that alternates $x$'s and $y$'s. The image is then a word that alternates $a$'s and $b$'s and no such word represents $e \in D_\infty$ by Proposition 2.1.

Given two groups $A$ and $B$ there are "coordinate inclusions" mapping the groups $A$ and $B$ into $A \oplus B$:

$$\phi_A(a) = (a, e_B) \text{ and } \phi_B(b) = (e_A, b).$$

Let $\phi : A * B \twoheadrightarrow A \oplus B$ be the map induced by these coordinate inclusions. As an example, if $g = a_1 b_1 a_2 b_2 a_3 \in A * B$ then

$$\phi(g) = (a_1 a_2 a_3, b_1 b_2) \in A \oplus B.$$

This map cannot be injective (if $A$ and $B$ are non-trivial). Take any two non-identity elements $a \in Z \setminus \{e_A\}$ and $b \in B \setminus \{e_B\}$, and form their *commutator*, which is defined as

$$[a, b] = a^{-1} b^{-1} a b.$$

The word $a^{-1} b^{-1} a b$ is a reduced word, hence $a^{-1} b^{-1} a b \neq e \in A * B$. However, $\phi([a, b]) = (a^{-1} a, b^{-1} b) = e \in A \oplus B$. Thus the kernel of $\phi$ contains more than the identity element.

**Theorem 3.35.** *Let $\phi : A * B \twoheadrightarrow A \oplus B$ be the map induced by the coordinate inclusions of $A$ and $B$ into $A \oplus B$. The kernel of $\phi$ is generated by the set of commutators of non-identity elements of $A$ and $B$*

$$S = \{[a, b] \mid a \in A \setminus \{e_A\} \text{ and } b \in B \setminus \{e_B\}\}$$

*and this kernel is a free group with basis $S$.*

**Example 3.36.** In the case of $D_\infty \approx \mathbb{Z}_2 * \mathbb{Z}_2$ the map $\phi$ is onto $\mathbb{Z}_2 \oplus \mathbb{Z}_2$ and by inspection one sees that $g \in D_\infty$ is in the kernel if and only if the reduced expression for $g$ has an even number of $a$'s and an even number of $b$'s. From this, it is not too difficult to believe that the kernel is generated by the element $[a, b] = abab$, and therefore the kernel is isomorphic to $\mathbb{Z}$.

We now present a proof of Theorem 3.35 based on the definition of a free group; a geometric argument is outlined in Section 3.8.

**Lemma 3.37.** *Let $K$ denote the subgroup of $A * B$ generated by $S$. Then $K$ is a normal subgroup.*

*Proof.* Let $\alpha \in A$ and $[a, b]$ be a commutator in $S$. Then

$$
\begin{aligned}
\alpha[a, b]\alpha^{-1} &= \alpha a^{-1}b^{-1}ab\alpha^{-1} \\
&= \alpha a^{-1}b^{-1}(a\alpha^{-1}b \cdot b^{-1}\alpha a^{-1})ab\alpha^{-1} \\
&= (\alpha a^{-1}b^{-1}a\alpha^{-1}b)(b^{-1}\alpha a^{-1}ab\alpha^{-1}) \\
&= (\alpha a^{-1}b^{-1}a\alpha^{-1}b)(b^{-1}\alpha b\alpha^{-1}) \\
&= [a\alpha^{-1}, b][\alpha^{-1}, b]^{-1}.
\end{aligned}
$$

Thus conjugating a commutator by any $\alpha \in A$ results in a product of a commutator and the inverse of a commutator, which is in $K$. A similar string of equations shows that conjugating a commutator by $\beta \in B$ results in an element of $K$:

$$
\beta[a, b]\beta^{-1} = [a, \beta^{-1}]^{-1}[a, b\beta^{-1}].
$$

Since $A$ and $B$ generate $A * B$, an induction argument shows that $gKg^{-1} < K$ for any $g \in A * B$. $\qquad\square$

**Lemma 3.38.** *The kernel of $\phi$ is the subgroup $K$ generated by $S$.*

We leave the proof of this lemma as an exercise:

**Exercise 3.39.** Let $\overline{A}$ be the subgroup of $A * B/K$ consisting of cosets of the form $\{aK \mid a \in A\}$, and let $\overline{B} = \{bK \mid b \in B\}$.

a. Show that $\overline{A} \cdot \overline{B} = A * B/K$.
b. Show that $\overline{A} \cap \overline{B} = eK$.

It follows from the definition of a direct sum that $A * B/K \approx \overline{A} \oplus \overline{B}$. You have established Lemma 3.38 once you have proved:

c. $\overline{A} \approx A$ and $\overline{B} \approx B$.

**Lemma 3.40.** *The subgroup $K$ is a free group with basis $S$.*

*Proof.* Our proof is by induction on the length of a reduced word in $\{S \cup S^{-1}\}^*$. Let

$$
\omega = [a_1, b_1]^{\epsilon_1}[a_2, b_2]^{\epsilon_2} \cdots [a_n, b_n]^{\epsilon_n}
$$

where $\epsilon_i = \pm 1$ and no commutator is listed next to its inverse. Thought of as an element in $\{A \cup B\}^*$, this word may not be reduced, but we can show it is in the same equivalence class as a reduced word $\varrho(\omega)$ whose length is at least $n+3$. Thus $\omega \neq e \in A * B$ and therefore $K$ is free with basis $S$.

Our proof establishes slightly more than is claimed above. Our induction hypothesis is: $\omega$ *is equivalent to a reduced word of length* $\geq n+3$ *and the reduced expression* $\varrho(\omega)$ *ends with* $a_n b_n$ *if* $\epsilon_n = +1$ *and* $b_n a_n$ *if* $\epsilon_n = -1$. The base case is immediate.

We assume that any reduced word in $\{S \cup S^{-1}\}^*$ of length $\leq n$ satisfies the induction hypothesis, and consider

$$\omega = [a_1, b_1]^{\epsilon_1} [a_2, b_2]^{\epsilon_2} \cdots [a_n, b_n]^{\epsilon_n} [a_{n+1}, b_{n+1}]^{\epsilon_{n+1}} .$$

With only a small loss of generality, we assume that $\epsilon_n = +1$. Hence $\omega$ is equivalent to the word

$$x_1 \cdots x_k a_n b_n [a_{n+1}, b_{n+1}]^{\epsilon_{n+1}}$$

with $x_1 \cdots x_k a_n b_n$ a reduced word of length $\geq n+3$.

If $\epsilon_{n+1} = +1$, then we may take

$$\varrho(\omega) = x_1 \cdots x_k a_n b_n a_{n+1}^{-1} b_{n+1}^{-1} a_{n+1} b_{n+1},$$

which is reduced, of length $\geq n+7$, and ends with the correct letters. If $\epsilon_{n+1} = -1$ then

$$\omega \sim x_1 \cdots x_k a_n b_n b_{n+1}^{-1} a_{n+1}^{-1} b_{n+1} a_{n+1} .$$

If $b_n \neq b_{n+1}$ we can let $b = b_n b_{n+1}^{-1} \in B \setminus \{e_B\}$ and then

$$\varrho(\omega) = x_1 \cdots x_k a_n b a_{n+1}^{-1} b_{n+1} a_{n+1},$$

which is reduced, of length $\geq n+6$, and ends with the correct letters. Finally, if $b_n = b_{n+1}$, then $a_n \neq a_{n+1}$, as $\omega$ was freely reduced when thought of as a product of commutators. We may set

$$a = a_n a_{n+1}^{-1} \in A \setminus \{e_A\}$$

and then

$$\varrho(\omega) = x_1 \cdots x_k a b_{n+1} a_{n+1},$$

which is reduced, of length $\geq n+4 = (n+1)+3$, and ends with the correct sequence of letters. $\qquad \square$

We can now establish the title of this section:

**Corollary 3.41.** *If $A$ and $B$ are finite, then $A * B$ contains a normal subgroup of index $|A| \times |B|$ that is free of rank $(|A| - 1) \times (|B| - 1)$.*

*Proof of Corollary 3.41.* The index of the kernel is the size of the quotient, which is $|A| \times |B|$. The size of $S$ is $(|A| - 1) \times (|B| - 1)$ by its definition. $\square$

**Example 3.42.** The kernel of $\mathbb{Z}_n * \mathbb{Z}_m \twoheadrightarrow \mathbb{Z}_n \oplus \mathbb{Z}_m$ is free of rank $(n - 1)(m - 1)$.

## 3.8 A Geometric View of Theorem 3.35

The proof of Theorem 3.35 given in Section 3.7 is algebraic. There is also a geometric argument which we outline here in the special case of $\mathbb{Z}_3 * \mathbb{Z}_4$. Recall our notation for the generators: the generator $a$ fixed a vertex of valence 4 and $b$ fixed a vertex of valence 3 in $T_{3,4}$. We will show:

**Proposition 3.43.** *Let $\phi : \mathbb{Z}_3 * \mathbb{Z}_4 \twoheadrightarrow \mathbb{Z}_3 \oplus \mathbb{Z}_4$ be the map induced by the "coordinate" inclusions of $\mathbb{Z}_3$ and $\mathbb{Z}_4$ into $\mathbb{Z}_3 \oplus \mathbb{Z}_4$. The kernel of $\phi$ is generated by the set of commutators of non-identity elements of $\mathbb{Z}_3$ and $\mathbb{Z}_4$*

$$S = \{[a, b] \mid a \in \mathbb{Z}_4 \setminus \{0\} \text{ and } b \in \mathbb{Z}_3 \setminus \{0\}\}$$

*and this kernel is a free group with basis $S$.*

The proof begins by noticing that $K = \text{Ker}(\phi)$ acts on $T_{3,4}$ since it is a subgroup of $\mathbb{Z}_3 * \mathbb{Z}_4$.

**Lemma 3.44.** *The kernel $K$ acts freely on $T_{3,4}$.*

*Proof.* We need to show that no non-identity element fixes any vertex or edge in $T_{3,4}$. If $e \in E(T_{3,4})$ then Exercise 3.31 shows that $\text{Stab}(e)$ is trivial. Corollary 3.30 shows that the vertex stabilizers are conjugates of $\mathbb{Z}_3$ and $\mathbb{Z}_4$. However, $\phi(g \cdot a^i \cdot g^{-1}) = (0, i) \in \mathbb{Z}_3 \oplus \mathbb{Z}_4$ for any $g \in A * B$. Similarly $\phi(g \cdot b^j \cdot g^{-1}) = (j, 0) \in \mathbb{Z}_3 \oplus \mathbb{Z}_4$. Thus the intersection of any vertex stabilizer with the kernel $K$ is just the identity element. Thus no element of the kernel $K$ stabilizes a vertex. It follows that the action $K \curvearrowright T_{3,4}$ is free. $\square$

Since $K$ acts freely on a tree, Theorem 3.20 implies the following result.

**Corollary 3.45.** *The kernel $K$ is a free group.*

Theorem 3.20 also gives a concrete way to find a basis for $K$ using a fundamental domain. In describing a fundamental domain for the action $K \curvearrowright \mathcal{T}_{3,4}$ we use the notation introduced in Section 3.5. In particular, $a$ is the generator of order 4 and $b$ is the generator of order 3. To construct a CORE for $K \curvearrowright \mathcal{T}_{3,4}$ we begin with the vertex we called "the origin" and we add edges and vertices, until we can no longer add vertices coming from distinct orbits.

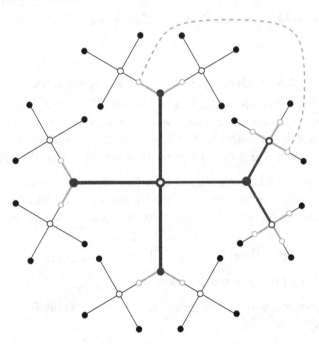

Fig. 3.14. A CORE and fundamental domain for the action of the kernel $K$ on $\mathcal{T}_{3,4}$. The dotted line indicates two half-edges that are connected by the action of $K$.

One possible CORE is shown in Figure 3.14. At first glance, our CORE may appear to be too small. All of the vertices in CORE are contained in $a^i \cdot e_e$ or $b^j \cdot e_e$, where $e_e$ is the edge joining the vertex fixed by $a$ to the vertex fixed by $b$. The vertex labelled NL in Figure 3.12 is an end of $a^3 b \cdot e_e$, and the image of $a^3 b$ in $\mathbb{Z}_3 \oplus \mathbb{Z}_4$ is not the same as the image of a single power of $a$ or $b$. However, the vertex labelled NL is also an end of $a^3 ba \cdot e_e$, and the image of $a^3 ba$ is the same as the image of $b$ in the quotient. One further check that the CORE, and the corresponding fundamental domain formed by adding half-edges, is correct comes from counting the vertices of each type. Because the index of $\mathbb{Z}_3$ in $\mathbb{Z}_3 \oplus \mathbb{Z}_4$ is four, there must be exactly four vertices of valence 3 in CORE. Similarly there need to be exactly three vertices of valence 4.

To find the basis associated with our fundamental domain, we need to see how the half-edges pair up. The half-edge connecting the fundamental domain to the vertex labelled NL comes from the element $a^3b \in \mathbb{Z}_3 * \mathbb{Z}_4$. The half-edge connecting the fundamental domain to the vertex labelled ELE comes from the element $ba^3$. But in the quotient group, $\mathbb{Z}_3 \oplus \mathbb{Z}_4$, $a^3b = ba^3$, and so these half-edges are paired. The corresponding element of the basis for the kernel $K$ is the commutator $[b, a^3]$. Continuing in this manner one gets the basis claimed in Proposition 3.43.

## 3.9 Finite Groups Acting on Trees

Given any group $G$, there is a tree $\mathcal{T}$, and a non-trivial action of $G$ on $\mathcal{T}$. One way to construct such an action on a tree is by having the vertex set of $\mathcal{T}$ correspond to the elements of $G$, with an additional distinct vertex "$\star$." For each $g \in G$ there is an edge joining $\star$ to the vertex associated with $g$, and there are no other edges. The action is induced by declaring that $g \in G$ takes the vertex associated to $h$ to the vertex associated to $gh$, and it takes $\star$ to $\star$. With this definition, we have constructed an injective map from $G$ to $\text{SYM}(\mathcal{T})$, hence the claim that the action $G \curvearrowright \mathcal{T}$ is "non-trivial." However, there is a sense in which this action is trivial: there is a point that is never moved by the action of $G$: $\text{Stab}(\star) = G$.

Jean-Pierre Serre defined a group $G$ to have *Property FA* if, whenever there is an action of $G$ on a tree $\mathcal{T}$, there is a point $x \in \mathcal{T}$ such that $\text{Stab}(x) = G$. (The letter "F" corresponds to "fixed" while the "A" comes from the French word for tree: *arbre*.)

**Theorem 3.46.** *All finite groups have Serre's Property FA.*

*Proof.* Let $G$ be finite and let $G \curvearrowright \mathcal{T}$ be an action of $G$ on a tree. Let $v$ be any vertex of $\mathcal{T}$ and consider the orbit of $v$ under the action of $G$:

$$\mathcal{O}(v) = \{g \cdot v \mid g \in G\} .$$

Let $\mathcal{T}_{\mathcal{O}(v)}$ be the subtree created by taking the union of all the vertices and edges of $\mathcal{T}$ that occur in minimal-length paths connecting elements of $\mathcal{O}(v)$. Since $g \cdot \mathcal{O}(v) = \mathcal{O}(v)$ for any $g \in G$, it follows that $G$ acts on $\mathcal{T}_{\mathcal{O}(v)}$.

If we have been lucky, and $\mathcal{O}(v) = v$, we have our fixed point. Otherwise $\mathcal{T}_{\mathcal{O}(v)}$ contains vertices of valence 1. (Vertices of valence 1 are called *leaves* of the tree.) The action of $G$ must permute the leaves of $T_{\mathcal{O}(v)}$, hence it also permutes all the non-leaf vertices of $T_{\mathcal{O}(v)}$. Let $T'$ be the tree formed by removing the leaves of $T_{\mathcal{O}(v)}$ and the edges incident to these leaves. We now have an action of $G$ on $T'$. If $T'$ is not an edge or a single vertex, then we can remove the leaves of $T'$ to form a smaller subtree $T''$ on which $G$ acts.

Because $G$ is finite, $\mathcal{O}(v)$ and $T_{\mathcal{O}(v)}$ are finite as well. Each time we trim off leaves, we create even smaller subtrees, and so by continuing in this manner we eventually reduce our original subtree to either a single vertex or a single edge. If it is an edge, then either all of the elements of $G$ fix the edge, or some elements of $G$ invert the edge. In either case, the center point of the edge is fixed. $\qquad\square$

Examining the proof of Theorem 3.46 we see that $G$ either fixes a vertex of $T$ or there is an edge $e$ whose midpoint is fixed by the action of $G$. If $T$ is bipartite and $G$ preserves the bipartite structure, then this second possibility cannot happen, and so we get:

**Corollary 3.47.** *If $T$ is a bipartite tree, and $G$ is a finite group contained in $\mathrm{SYM}^+(T)$, then $G$ fixes a vertex.*

**Corollary 3.48.** *If $H$ is a finite subgroup of $A * B$ then $H$ is conjugate to a subgroup of $A$ or $B$.*

*Proof.* The vertex stabilizers are conjugates of $A$ and $B$ for the action of $A * B$ on its associated tree, by Corollary 3.30. By Corollary 3.47, a finite subgroup of $A * B$ must fix a vertex, hence $H < gAg^{-1}$ or $gBg^{-1}$ for some $g \in A * B$. $\qquad\square$

## 3.10 Serre's Property FA and Infinite Groups

It is not just the finite groups that have Serre's Property FA. In fact, a number of interesting infinite groups are also unable to act on a tree without a global fixed point. Before we can present a few examples of such groups, we need to establish a small amount of background material.

**Definition 3.49.** Let $G$ act on $X$. Then for any element $g \in G$ define the *fixed subset* to be

$$\mathrm{Fix}(g) = \{x \in X \mid g \cdot x = x\}.$$

If $H$ is any subgroup of $G$ then set

$$\mathrm{Fix}(H) = \bigcap_{h \in H} \mathrm{Fix}(h).$$

Serre's definition of Property FA used the notion of fixed subsets, although his notation was different from that given above. If $G \curvearrowright X$ then he used $X^G$ to denote $\mathrm{Fix}(G)$, as in the following definition from [Se77]:

Nous nous intéressons aux groupes $G$ ayant la propriété: (FA) – Quel que soit l'arbre $X$ sur lequel opère $G$, on a $X^G \neq \emptyset$.

**Lemma 3.50.** *Let $G$ act on a tree $T$. Then either $\mathrm{Fix}(G)$ is empty, or it is a subtree of $T$.*

*Proof.* Assume $\mathrm{Fix}(G)$ is not empty. If $v$ and $w$ are two distinct fixed points in $\mathrm{Fix}(G)$, then there is exactly one minimal-length edge path containing $v$ and $w$. Since this edge path is unique, it must also be fixed (pointwise) by $G$. Thus $\mathrm{Fix}(G)$ is a connected subgraph of $T$, hence it is a subtree. $\square$

If $S$ is a subset of $G$ we have used $\langle S \rangle$ to denote the subgroup of $G$ generated by the set $S$. In the following lemma we use this notation in the context of two subgroups, $H, K < G$. Namely, $\langle H, K \rangle$ is the subgroup of $G$ generated by the elements in the set $H \cup K$. In particular, the elements of $\langle H, K \rangle$ consist of all possible products of finitely many elements from $H$ and $K$.

**Lemma 3.51.** *Let $G$ act on a tree $T$ and let $H$ and $K$ be two subgroups of $G$. Then*

$$\mathrm{Fix}(\langle H, K \rangle) = \mathrm{Fix}(H) \cap \mathrm{Fix}(K).$$

*Proof.* If $x \in \mathrm{Fix}(H) \cap \mathrm{Fix}(K)$, then $x$ is fixed by every $h \in H$ and $k \in K$, hence $x$ is fixed by every element of $\langle H, K \rangle$. Thus we have $\mathrm{Fix}(\langle H, K \rangle) \supset \mathrm{Fix}(H) \cap \mathrm{Fix}(K)$. To establish the other containment, notice that, because $H$ and $K$ are subgroups of $\langle H, K \rangle$, it follows that $\mathrm{Fix}(\langle H, K \rangle) \subset \mathrm{Fix}(H)$ and $\mathrm{Fix}(\langle H, K \rangle) \subset \mathrm{Fix}(K)$. Thus we have the other containment: $\mathrm{Fix}(\langle H, K \rangle) \subset \mathrm{Fix}(H) \cap \mathrm{Fix}(K)$. $\square$

**Proposition 3.52** (Triangle Condition). *Let $G$ be generated by three subgroups, $H_1, H_2$ and $H_3$. Define $H_{ij}$ to be $\langle H_i, H_k \rangle$. If all of the $H_i$ and $H_{ij}$ are finite, then $G$ has Serre's Property FA.*

*Proof.* We begin by showing that

$$\text{Fix}(H_1) \cap \text{Fix}(H_2) \cap \text{Fix}(H_3) \neq \emptyset.$$

Since each $H_i$ is assumed to be finite, $\text{Fix}(H_i) \neq \emptyset$ by Theorem 3.46. Further, because $\text{Fix}(H_i) \cap \text{Fix}(H_j) = \text{Fix}(H_{ij})$ (Lemma 3.51) and $\text{Fix}(H_{ij}) \neq \emptyset$ (because $H_{ij}$ is finite), we have

$$\text{Fix}(H_i) \cap \text{Fix}(H_j) \neq \emptyset.$$

We may then take $v$ to be a point in $\text{Fix}(H_1) \cap \text{Fix}(H_2)$ and $w$ to be a point in $\text{Fix}(H_2) \cap \text{Fix}(H_3)$. Thus $v$ and $w$, as well as the minimal-length edge path containing $v$ and $w$, are all contained in $\text{Fix}(H_2)$. However, $v$ and $w$ are also in $\text{Fix}(H_1) \cup \text{Fix}(H_3)$. As this union is a subtree, the minimal-length edge path containing $v$ and $w$ is also in the union $\text{Fix}(H_1) \cup \text{Fix}(H_3)$. It follows that this edge path must intersect $\text{Fix}(H_1) \cap \text{Fix}(H_3)$, implying that there is a point of intersection of $\text{Fix}(H_2)$ with $\text{Fix}(H_i) \cap \text{Fix}(H_j)$.

Essentially the same argument as is given for Lemma 3.51 shows that

$$\text{Fix}(\langle H_1, H_2, H_3 \rangle) = \text{Fix}(H_1) \cap \text{Fix}(H_2) \cap \text{Fix}(H_3).$$

The argument of the previous paragraph shows the right side is non-empty, while we have assumed that $G = \langle H_1, H_2, H_3 \rangle$. It follows that $\text{Fix}(G) \neq \emptyset$.                                                    $\square$

**Example 3.53.** In Chapter 2 we discussed groups generated by reflections. The groups $W_{236}, W_{244}$ and $W_{333}$ are all groups generated in the lines formed by certain Euclidean triangles. These three groups all have Property FA. To see why, consider $W_{333}$ and let $\{r, g, b\}$ be the three generating reflections. Set $H_1 = \langle r \rangle \approx \mathbb{Z}_2$, $H_2 = \langle g \rangle \approx \mathbb{Z}_2$, and $H_3 = \langle b \rangle \approx \mathbb{Z}_2$. Then each of the three groups $H_{ij}$ is isomorphic to $D_3$, the dihedral group of order 6. (The groups $H_{ij}$ are, in fact, the stabilizers of the vertices of the fundamental domain for the action of $W_{333}$ on $\mathbb{R}^2$.) Since all of these groups are finite, and a generating set of $W_{333}$ is contained in $H_1 \cup H_2 \cup H_3$, it follows from Proposition 3.52 that $W_{333}$ has Property FA.

The only change that occurs when considering $W_{236}$ and $W_{244}$ is that the $H_{ij}$ can equal $D_2, D_3, D_4$ or $D_6$, depending on the reflection group under consideration and which pair of generators is chosen.

For the rest of this section we will explore the group of automorphisms of a free group $\mathbb{F}_n$, with the goal of establishing that $\text{Aut}(\mathbb{F}_n)$, for $n \geq 3$, has Serre's property FA. In order not to take up too many pages, our exposition will be fairly terse. Any reader who finds this discussion interesting is encouraged to fill in the missing details, and to consult the various references mentioned.

Let $S = \{x_1, \ldots, x_n\}$ be a basis for $\mathbb{F}_n$ and let $\mathcal{S} = S \cup S^{-1}$ be the union of this basis and its inverses. There are a number of automorphisms of $\mathbb{F}_n$ that come from permutations of $\mathcal{S}$. For example, if $\sigma \in \mathrm{SYM}_n$ is a permutation of $\{1, \ldots, n\}$ then one can define an associated automorphism $\sigma \in \mathrm{Aut}(\mathbb{F}_n)$ induced by requiring

$$\sigma(x_i) = x_{\sigma(i)}$$

(where we are abusing the use of the symbol "$\sigma$"). To verify that this is indeed an automorphism, notice that $\sigma$ must be surjective, since the image of $S$ is $S$. Further, it must be injective, since freely reduced words are mapped to freely reduced words, hence no non-identity element is mapped to the identity. The subgroup of $\mathrm{Aut}(\mathbb{F}_n)$ generated by these *permutation automorphisms* is isomorphic to $\mathrm{SYM}_n$.

Similarly one can exchange generators and their inverses. That is, for each $i \in \{1, \ldots, n\}$ there is an automorphism $\iota_i \in \mathrm{Aut}(\mathbb{F}_n)$ induced by requiring

$$\iota_i(x_j) = \begin{cases} x_j^{-1} & \text{if } i = j \\ x_j & \text{otherwise.} \end{cases}$$

The subgroup generated by these *inversion automorphisms* is isomorphic to $(\mathbb{Z}_2)^n \approx \mathbb{Z}_2 \oplus \cdots \oplus \mathbb{Z}_2$. The subgroup of $\mathrm{Aut}(\mathbb{F}_n)$ generated by the permutation and inversion automorphisms injects into $\mathrm{SYM}(\mathcal{S})$, hence it must be finite. It is, in fact, a group of order $2^n \cdot n!$. We denote this subgroup of $\mathrm{Aut}(\mathbb{F}_n)$ by $\Omega_n$.

Jakob Nielsen introduced another important collection of automorphisms, which we call *Nielsen automorphisms*. These are the automorphisms $\{\rho_{ij}\}$ induced by

$$\rho_{ij}(x_k) = \begin{cases} x_i x_j & \text{if } k = i \\ x_k & \text{otherwise.} \end{cases}$$

That this defines a homomorphism from $\mathbb{F}_n$ to $\mathbb{F}_n$ is an application of Theorem 3.15. To see that it is a bijection, one needs to establish that the image of $S = \{x_1, \ldots, x_n\}$ under $\rho_{ij}$ is a basis for $\mathbb{F}_n$. This is essentially part (a) of Exercise 6.

The following theorem (first proved by Nielsen in 1924 [Ni24]) is not too difficult, and the interested reader can find a good exposition of it in [MKS66]. The details, however, take too many pages for us to include here.

**Theorem 3.54.** *The elementary automorphisms generate* $\mathrm{Aut}(\mathbb{F}_n)$.

Having a generating set allows us to begin an argument that establishes the Triangle Condition (3.52) for $\mathrm{Aut}(\mathbb{F}_n)$. Because the Triangle Condition requires that the group is generated by finite subgroups, and the generators $\rho_{ij}$ have infinite order, our initial task will be to transform our generating set into one that does not use the $\rho_{ij}$. The first step is to notice that if $\sigma$ is a permutation automorphism then $\sigma\rho_{ij}\sigma^{-1} = \rho_{\sigma(i)\sigma(j)}$. From this, one gets:

**Corollary 3.55.** *The automorphism group of a free group,* $\mathrm{Aut}(\mathbb{F}_n)$, *is generated by* $\Omega_n$ *and* $\rho_{12}$.

In order to remove $\rho_{12}$ from our generating set, we introduce three automorphisms of $\mathbb{F}_n$, each of which has order 2. The first is $\theta$, which takes $x_1$ to $x_1 x_2$ and $x_2$ to $x_2^{-1}$, leaving all the other elements in the basis fixed. This automorphism can be expressed as a product of elementary automorphisms: $\theta = \rho_{12}\iota_2$. To see that the order of $\theta$ is 2, first notice that $\theta^2(x_i) = x_i$ for all $i \geq 2$, and then compute

$$\theta^2(x_1) = \theta(x_1 x_2) = \theta(x_1)\theta(x_2) = x_1 x_2 \cdot x_2^{-1} = x_1.$$

The remaining two automorphisms are particular elements of $\Omega_n$. Define $\tau \in \mathrm{Aut}(\mathbb{F}_n)$ by $\tau(x_1) = x_1^{-1}$, $\tau(x_2) = x_3$, $\tau(x_3) = x_2$ and $\tau(x_i) = x_i$ for $i \geq 3$. (So $\tau$ is a product of an inversion and a permutation automorphism.) Finally, let $\eta$ be defined by $\eta(x_1) = x_2^{-1}$, $\eta(x_2) = x_1^{-1}$, and $\eta(x_i) = x_i$ for $i \geq 3$.

Let $\mathrm{SYM}_{n-2}$ be the subgroup of $\mathrm{Aut}(\mathbb{F}_n)$ generated by permutation automorphisms where $\sigma(1) = 1$ and $\sigma(2) = 2$, and let $H_1$ be the subgroup generated by $\mathrm{SYM}_{n-2}$ along with $\iota_n$ and $\eta$. As all of these automorphisms are contained in $\Omega_n$, $H_1$ is a subgroup of $\Omega_n$, hence $H_1$ is a finite group. Define $H_2$ to be the cyclic subgroup generated by $\theta$, and let $H_3$ be the cyclic subgroup generated by $\tau$.

**Lemma 3.56.** *For* $n \geq 3$, *the subgroups* $\{H_1, H_2, H_3\}$ *of* $\mathrm{Aut}(\mathbb{F}_n)$ *satisfy the Triangle Condition.*

*Proof.* First, we have to show that $\{H_1 \cup H_2 \cup H_3\}$ is a generating set for $\mathrm{Aut}(\mathbb{F}_n)$. Let $\sigma_{ij}$ denote the permutation automorphism that exchanges $x_i$ and $x_j$. Conjugating the automorphism $\sigma_{3n} \in H_1$ by $\tau \in H_3$ we get

$\sigma_{2n}$, as can be seen by the following computations:

$$
\begin{array}{ccccccc}
x_1 & \xrightarrow{\tau} & x_1^{-1} & \xrightarrow{\sigma_{3n}} & x_1^{-1} & \xrightarrow{\tau} & x_1 \\
x_2 & \xrightarrow{\tau} & x_3 & \xrightarrow{\sigma_{3n}} & x_n & \xrightarrow{\tau} & x_n \\
x_3 & \xrightarrow{\tau} & x_2 & \xrightarrow{\sigma_{3n}} & x_2 & \xrightarrow{\tau} & x_3 \\
x_n & \xrightarrow{\tau} & x_n & \xrightarrow{\sigma_{3n}} & x_3 & \xrightarrow{\tau} & x_2
\end{array}
$$

(Every other $x_i$ is fixed by both $\tau$ and $\sigma_{3n}$.) Conjugating $\iota_n \in H_1$ by $\sigma_{2n}$ results in $\iota_2$; conjugating $\sigma_{3n}$ by $\sigma_{2n}$ creates $\sigma_{23}$. We then also have $\iota_1$, as $\iota_1 = \sigma_{23}\tau$, and $\sigma_{12}$, which is $\eta\iota_1\epsilon_2$. Thus the subgroup generated by the $H_i$ contains all the "transposition automorphisms" $\sigma_{ij}$ and all the inversions $\iota_i$, hence it contains $\Omega_n$. Since $\theta \in H_2$, and $\iota_2 \in \Omega_n$, it follows that $\rho_{12} = \theta\iota_2 \in \langle H_1, H_2, H_3 \rangle$. Thus $\langle H_1, H_2, H_3 \rangle$ contains $\Omega_n$ and $\rho_{12}$. Thus, by Corollary 3.55, $\langle H_1, H_2, H_3 \rangle = \mathrm{Aut}(\mathbb{F}_n)$.

As we already know each $H_i$ is finite, it remains to verify that each $H_{ij} = \langle H_i, H_j \rangle$ is finite. This we do, case by case.

The group $H_{13}$ is generated by elements of $\Omega_n$, hence $H_{13}$ is a subgroup of $\Omega_n$, and is therefore finite.

The group $H_{23}$ is generated by two elements of order 2, $\theta$ and $\tau$. Their product is the automorphism that takes $x_1$ to $x_2^{-1}x_1^{-1}$, $x_2$ to $x_3$ and $x_3$ to $x_2^{-1}$. Direct computation then shows that $\theta\tau$ has order 4, and therefore $H_{23} \approx D_4$.

The subgroup generated by $\mathrm{SYM}_{n-2}$ and $\iota_n$ is $\Omega_{n-2}$, that is, it consists of all elements of $\Omega_n$ that fix $x_1$ and $x_2$. The subgroup $H_{12}$ is generated by this $\Omega_{n-2}$ and the elements $\theta$ and $\eta$. The elements of $\Omega_{n-2}$ commute with both $\theta$ and $\eta$ so $H_{12} \approx \Omega_{n-2} \oplus \langle \theta, \eta \rangle$. Further, $\theta$ and $\eta$ are both of order 2, and their product $\theta\eta$ has order 3. Thus $\langle \theta, \eta \rangle \approx D_3$, and $H_{12}$ is therefore the finite group $\Omega_{n-2} \oplus D_3$. $\qquad\square$

Lemma 3.56 and Proposition 3.52 imply the following result:

**Theorem 3.57.** *The automorphism group of a free group $\mathbb{F}_n$, for $n \geq 3$, has Serre's Property FA.*

This result was first established by Bogopolski in [Bo87]; the argument outlined here was given by Bridson in [Br08]. A third, also accessible, argument appears in a paper by Culler and Vogtmann [CV96].

The group $\mathrm{GL}_n(\mathbb{Z})$, the group of invertible $n \times n$ matrices with integer entries, is a quotient of $\mathrm{Aut}(\mathbb{F}_n)$. To see this, note that the image of a commutator under an automorphism $\alpha \in \mathrm{Aut}(\mathbb{F}_n)$ is another commutator. Thus any automorphism of $\mathbb{F}_n$ induces an automorphism of $\mathbb{Z}^n$, and $\mathrm{GL}_n(\mathbb{Z}) \approx \mathrm{Aut}(\mathbb{Z}^n)$.

**Lemma 3.58.** *If $G$ has Property FA, then any quotient of $G$ also has Property FA.*

*Proof.* Let $G/N \approx H$. If $H$ acts on a tree $\mathcal{T}$ without fixed point, then so does $G$. The action is given by $g \cdot \mathcal{T} = (gN) \cdot \mathcal{T}$.                □

**Corollary 3.59.** *For $n \geq 3$, $GL_n(\mathbb{Z})$ has Serre's Property FA.*

   J-P. Serre used a variation on the Triangle Condition to establish that $SL_n(\mathbb{Z})$ has Property FA for all $n \geq 3$. (A variation of the argument above also establishes this fact.) This runs somewhat counter to the intuition given by Proposition 3.7, which states that the matrix group $SL_2(\mathbb{Z})$ is virtually free. Given the fact that free groups are closely connected to trees, it is not surprising that the group $SL_2(\mathbb{Z})$ does not have Property FA, and in fact it admits a nice action on $\mathcal{T}_{2,3}$. Further, the larger group $GL_2(\mathbb{Z})$ also admits a nice action on $\mathcal{T}_{2,3}$, hence the restriction that $n \geq 3$ cannot be removed from Corollary 3.59 or Theorem 3.57. These actions are some of the more elementary examples of how a free product *with amalgamation* acts on a tree. See Serre's book ([Se77] or the English translation [Se80]) for an excellent discussion of this and related topics.


## Exercises

(1)   Define the distance between two vertices in a tree to be the number of edges that occur in the unique reduced path between them.

     a. Prove that the number of vertices a distance $n$ from a fixed vertex $v$ in $\mathcal{T}_m$ is $m \cdot (m-1)^{n-1}$.

     b. Find a formula for the number of vertices a distance $n$ from a fixed vertex $v$ in $\mathcal{T}_{\{m,n\}}$. (You will actually need to find two formulas, one for the black vertices and one for the white vertices.)

(2)   Prove that $\mathbb{Z}^n$ is a subgroup of Homeo($\mathbb{R}$), for all $n \in \mathbb{N}$.

(3)   In Remark 3.11 it is stated that the subgroup of Homeo($\mathbb{R}$) generated by $f(x) = x^3$ and $g(x) = x + 1$ is a free group.

     a. What are $f^{-1}$ and $g^{-1}$?

     b. Show that no freely reduced word in $\{f, g, f^{-1}, g^{-1}\}^*$, of length at most 3, represents the identity.

(4)   Prove the Ping Pong Lemma (Lemma 3.10).

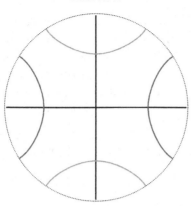

Fig. 3.15. The hyperbolic isometry whose axis is the horizontal line and the hyperbolic isometry whose axis is the vertical line generate a free subgroup of the isometry group of the hyperbolic plane.

(5) (This problem is for people who have studied hyperbolic geometry.) Consider the hyperbolic plane with six specified geodesics, as shown in Figure 3.15. Let $x$ be the hyperbolic isometry of the hyperbolic plane whose axis is the horizontal line, and which takes the geodesic on the left to the geodesic on the right. Let $y$ be the hyperbolic isometry with axis the vertical line, and which takes the lower geodesic to the upper geodesic. Show that $\{x, y\}$ generate a free group.

(6) Let $\mathbb{F}_2$ be the free group with basis $\{x, y\}$.

    a. Show that $\{x, xy\}$ is also a basis for $\mathbb{F}_2$.

    b. Show that $\{x, x^2 y\}$ is also a basis for $\mathbb{F}_2$.

    c. Show that there are infinitely many distinct bases of $\mathbb{F}_2$.

(7) Let $g \in \mathbb{F}_2$ and express $g$ as a word in the generators and their inverses: $g = a_1 a_2 \cdots a_n$. This word is said to be *cyclically reduced* if it is freely reduced and $a_1 \neq a_n^{-1}$. Let $\widehat{g}$ be the element produced by progressively removing such pairs of letters until the resulting expression is cyclically reduced.

    a. Show that $g$ is conjugate to $\widehat{g}$.

    b. Show that $g$ and $h$ are conjugate if and only if $\widehat{g}$ and $\widehat{h}$ are cyclic rearrangements of each other.

(8)  Let $\mathbb{F}_2$ be a free group with basis $\{x, y\}$. For any natural number $n$ let $\beta_n = \{y^{-i}xy^i \mid 1 \leq i \leq n\}$. Show that the subgroup of $\mathbb{F}_2$ generated by $\beta_n$ is free of rank $n$.

(9)  Show that $\mathbb{F}_2$ has exactly three distinct subgroups of index 2.

(10)  Let $\mathbb{F}_2$ be the free group with basis $\{x, y\}$. Define a map $\phi : \mathbb{F}_2 \twoheadrightarrow \mathbb{Z}^2$ by $\phi(x) = (1, 0)$ and $\phi(y) = (0, 1)$. Prove that the kernel of $\phi$ is a free group generated by

$$X = \{[x^k, y^l] \mid k, l \in \mathbb{Z} \text{ and } k \cdot l \neq 0\}.$$

Conclude that this kernel is a free group of infinite rank.

(11)  Show that $\mathbb{F}_n \not\approx \mathbb{F}_m$ if $n \neq m$.

(12)  Show that $\mathbb{F}_2 \approx \mathbb{Z} * \mathbb{Z}$.

(13)  Show that $A * B$ is a free group if and only if $A$ and $B$ are free.

(14)  Let $\mathbb{F}_2$ be a free group with basis $\{x, y\}$, and let $K$ be the kernel of the homomorphism $\phi : \mathbb{F}_2 \to \mathbb{Z}_3$ induced by $\phi(x) = \phi(y) = 1$.

　　a.  Find a fundamental domain for the action of $K$ on the Cayley graph of $\mathbb{F}_2$.

　　b.  Find a basis for the kernel $K$.

(15)  Let $\mathbb{F}_2$ be a free group of rank 2. Show that, for any $n \geq 2$, $\mathbb{F}_2$ contains a finite-index subgroup isomorphic to $\mathbb{F}_n$.

(16)  Let $N$ be an infinite-index normal subgroup of a free group $\mathbb{F}_n$ $(n \geq 2)$. Show that $N$ is not finitely generated.

(17)  Let $G = A * B$ and let $g = a_1 b_1 \cdots a_n b_n$ with the right-hand side being a reduced expression. Prove that the order of $g$ is infinite.

(18)  Let $A$ and $B$ be non-trivial groups. Prove that the center of $A * B$ is trivial.

(19)  Prove Theorem 3.33. You may want to use the proof of Theorem 3.15 as a model for your argument.

(20)  The alternating group $A_4$ is generated by $(12)(34)$ and $(123)$. Let $s$ generate a cyclic group of order 2, let $t$ generate a cyclic group of order 3, and let $\mathbb{Z}_2 * \mathbb{Z}_3$ be their free product. Finally, let $\phi : \mathbb{Z}_2 * \mathbb{Z}_3 \to A_4$ be the map induced by $s \mapsto (12)(34)$ and $t \mapsto (123)$.

　　a.  Show that $K = \text{Ker}(\phi)$ is a free group.

　　b.  Find a basis for $K$.

　　c.  If you are ambitious, replace "3" by "$n$."

(21)  The *centralizer* of an element $g \in G$, $C(g)$, is the subgroup of $G$ consisting of all elements that commute with $g$. Let $g \in A * B$.

　　a.  Show that if $g$ is in a conjugate of $A$ or $B$, then $C(g)$ is a subgroup of the same conjugate of $A$ or $B$.

b. Show that if $g$ is not in a conjugate of $A$ or $B$, then $C(g)$ is infinite cyclic.

(22) Let $G$ be the free product of two finite groups, $G \approx A * B$. Prove that if $H$ is a subgroup of $G$ that contains no elements of finite order (aside from the identity) then $H$ is a free group.

(23) In discussing free products we concentrated exclusively on the free product of two groups. One can also define free products of three (or more) groups. Prove that $(A * B) * C \approx A * (B * C)$, hence we may write $A * B * C$ to unambiguously denote the free product of three groups.

(24) Describe the Cayley graph of $\mathbb{Z}_2 * \mathbb{Z}_3 * \mathbb{Z}_4$, with respect to a set of cyclic generators of the factors.

(25) Let $\{x, y\}$ be a basis for $\mathbb{F}_2$ and let $\rho \in \mathrm{Aut}(\mathbb{F}_2)$ be the Nielsen automorphism induced by $\rho(x) = xy$ and $\rho(y) = y$. Describe $\rho^{-1}$.

# 4
# Baumslag–Solitar Groups

The first time I saw the group $\langle a, b \mid ab = b^2 a \rangle$ was in a paper of Graham Higman's around 1950, in the *Journal of the London Mathematical Society*. Actually the group may have been known to others. Anyway, the reference to Graham Higman is the reason I often complain about this group being attributed to Donald and myself.

–Gilbert Baumslag

Many of the infinite groups we have considered so far can be viewed as groups acting on the real line. The infinite dihedral group is one example and in Section 3.1.3 we showed that free groups also act on the real line. Let us now consider the subgroup of Homeo($\mathbb{R}$) generated by the linear functions $a$ and $b$, where $a(x) = 2x$ and $b(x) = x + 1$, and where the binary operation is function composition. The product $ab$ is then the function

$$x \xrightarrow{b} x + 1 \xrightarrow{a} 2x + 2$$

hence $ab(x) = 2x + 2$. One can also see that $ba(x) = 2x + 1$ since

$$x \xrightarrow{a} 2x \xrightarrow{b} 2x + 1 .$$

Thus $ab \neq ba$, so this group is not abelian. However, $b^2 a(x) = 2x + 2$, so $ab = b^2 a$, and therefore this group is not a free group with basis $\{a, b\}$. This group is referred to as a *Baumslag–Solitar* group.[1] In slightly greater generality, if $a(x) = nx$ for some integer $n \geq 2$, and $b$ is again $b(x) = x+1$, then the group generated by $\{a, b\}$ is the Baumslag–Solitar group denoted BS$(1, n)$.

---

[1] This terminology comes from the fact that Gilbert Baumslag and Donald Solitar wrote a key paper on the groups BS$(1, n)$ in 1962 [BS62]. As the quotation at the start of this chapter indicates, Baumslag has waged a vigorous, sustained, and ultimately doomed campaign against referring to BS$(1, 2)$ as a Baumslag–Solitar group. The paper by Graham Higman is actually from 1951 [Hi51].

The family of Baumslag–Solitar groups are most commonly defined in terms of generators and relations (see Section 3.3). It is a fact that the group $\mathrm{BS}(1,2)$ is the group with presentation $\langle a, b \mid ab = b^2 a \rangle$. In general, if $m$ and $n$ are positive integers, then the Baumslag–Solitar group $\mathrm{BS}(m,n)$ is defined to be the group given by the presentation $\langle a, b \mid ab^m = b^n a \rangle$.

The action of $\mathrm{BS}(1,2)$ on $\mathbb{R}$ is different than most actions we have previously considered. If $X$ is a geometric or topological object, like the real line, then an action of a finitely generated group $G \curvearrowright X$ is *discrete* if every orbit forms a discrete subset of $X$. That is, for each $x \in X$ there is a small neighborhood $N_x \ni x$ such that

$$\mathrm{Orb}(x) \cap N_x = x.$$

In the case of a group $G$ acting on the real line, one can rephrase this, to say that $G \curvearrowright \mathbb{R}$ is a *discrete* action if for every $x \in \mathbb{R}$ there is an $\epsilon > 0$ such that

$$\mathrm{Orb}(x) \cap (x - \epsilon, x + \epsilon) = x.$$

The action $D_\infty \curvearrowright \mathbb{R}$ described in Chapter 2 is an example of a discrete action on $\mathbb{R}$. The closed unit interval, $[0, 1]$, is a fundamental domain for $D_\infty \curvearrowright \mathbb{R}$, hence if $x \in \mathbb{R}$ then there is some $r \in [0, 1]$ and some $g \in D_\infty$ such that $g \cdot r = x$. It follows that the orbit of any $x \in \mathbb{R}$ has the form: $\{2n \pm r \mid n \in \mathbb{Z} \text{ and } r \in [0, 1]\}$. Thus, if $x$ is an integer, then set $\epsilon = 1/2$; otherwise set $\epsilon = \min\{|n - x|/2 \mid n \in \mathbb{Z}\}$. Then for any $x \in \mathbb{R}$ one has $\mathrm{Orb}(x) \cap (x - \epsilon, x + \epsilon) = x$, and so this action is discrete.

A discrete action of a finitely generated group on a space $X$ is *proper* if, for any $x \in X$, $\mathrm{Stab}(x)$ is finite. Again, the action $D_\infty \curvearrowright \mathbb{R}$ is proper, as, for any $x \in \mathbb{R}$,

$$\mathrm{Stab}(x) \approx \begin{cases} \mathbb{Z}_2 & \text{if } x \in \mathbb{Z} \\ e & \text{otherwise.} \end{cases}$$

Unlike the action $D_\infty \curvearrowright \mathbb{R}$, the action $\mathrm{BS}(1,2) \curvearrowright \mathbb{R}$ is neither discrete nor proper. To see that the action is not proper, notice that the stabilizer of $0 \in \mathbb{R}$ contains the subgroup generated by $a$, and $\langle a \rangle \approx \mathbb{Z}$. (In fact, Exercise 3 asks you to prove that $\mathrm{Stab}(0) = \langle a \rangle \approx \mathbb{Z}$.)

To show that the action $\mathrm{BS}(1,2) \curvearrowright \mathbb{R}$ is not discrete, notice that if $n$ is any positive integer then $a^{-n} b(0) = \dfrac{1}{2^n}$. Thus, given any $\epsilon > 0$, the orbit of $0 \in \mathbb{R}$ under the action of $\mathrm{BS}(1,2)$ intersects $(-\epsilon, \epsilon)$ in infinitely many points.

The fact that the orbit of 0 includes $\dfrac{1}{2^n}$ is a hint that to understand BS$(1,2)$ we should focus on the dyadic rationals. The *dyadic rationals* are the rational numbers that can be expressed as an integer over a power of 2. This set is denoted $\mathbb{Z}[1/2]$, and just to be explicit we repeat:

$$\mathbb{Z}[1/2] = \left\{ \frac{m}{2^n} \mid m \in \mathbb{Z} \text{ and } n \in \mathbb{N} \right\}.$$

If $m$ is any integer, then $a^n b^m (0) = \dfrac{m}{2^n}$, which is to say that the orbit of 0 under the action of BS$(1,2)$ contains $\mathbb{Z}[1/2]$. But each generator and its inverse maps $\mathbb{Z}[1/2]$ to $\mathbb{Z}[1/2]$:

$$a\left(\frac{m}{2^n}\right) = \frac{m}{2^{n-1}},$$
$$a^{-1}\left(\frac{m}{2^n}\right) = \frac{m}{2^{n+1}},$$
$$b\left(\frac{m}{2^n}\right) = \frac{m}{2^n} + 1 = \frac{m + 2^n}{2^n},$$
$$b^{-1}\left(\frac{m}{2^n}\right) = \frac{m}{2^n} - 1 = \frac{m - 2^n}{2^n}.$$

Thus the orbit of 0, or in fact any element of $\mathbb{Z}[1/2]$, is $\mathbb{Z}[1/2]$. This fact also follows from the following description of this group.

**Proposition 4.1.** *The elements of* BS$(1,2)$ *are the linear functions* $g : \mathbb{R} \to \mathbb{R}$ *that can be expressed in the form* $g(x) = 2^n \cdot x + \dfrac{m}{2^k}$ *where* $n, m$ *and* $k$ *are integers.*

*Outline of the proof.* We ask the reader to verify the following steps as a useful exercise.

1. Since BS$(1,2)$ is generated by linear functions, and the composition of linear functions is linear, every $g \in$ BS$(1,2)$ has the form $g(x) = \alpha x + \beta$.
2. Since BS$(1,2)$ is defined as the group generated by $a$ and $b$, every element can be expressed as a product of $a$'s, $b$'s, $a^{-1}$'s and $b^{-1}$'s. Arguing by induction on the length of these products one sees that if $g(x) = \alpha x + \beta$ then $\alpha = 2^n$ for some integer $n$ and $\beta \in \mathbb{Z}[1/2]$.
3. The translation $x \mapsto x + \frac{m}{2^k}$ can be expressed using our generators $a$ and $b$ as $a^{-k} b^m a^k$.
4. It follows that if
$$g(x) = 2^n \cdot x + \frac{m}{2^k}$$

then

$$g(x) = \left(a^{-k}b^m a^k\right) a^n.$$

Hence $g(x) = a^{-k}b^m a^{k+n}$ and therefore $\mathrm{BS}(1,2)$ is composed entirely of linear functions of the prescribed form.

$\square$

By its construction we know that $\mathrm{BS}(1,2)$ acts on the real line. But this action is neither discrete nor proper, and the tools developed in previous chapters seem to rely on, or at least prefer, discrete and proper actions. For example, the Drawing Trick (Remark 1.49) provides a method for constructing a Cayley graph of a group $G$, by using the orbit of a point which is moved freely by $G$ as the vertices of the Cayley graph. This method worked nicely in Chapter 2, where we used it to draw Cayley graphs of some groups generated by reflections, but it does not work well for the action $\mathrm{BS}(1,2) \curvearrowright \mathbb{R}$. Because of this, we delay discussing the Cayley graph of $\mathrm{BS}(1,2)$ with respect to $\{a,b\}$ until Section 5.4. Instead, we introduce an alternative set on which $\mathrm{BS}(1,2)$ acts.

**Corollary 4.2.** *The orbit of the closed interval* $[0,1]$ *under the action* $\mathrm{BS}(1,2) \curvearrowright \mathbb{R}$ *is the set of closed intervals*

$$\mathcal{I} = \{[x, x+2^i] \mid x \in \mathbb{Z}[1/2] \text{ and } i \in \mathbb{Z}\}.$$

*Further, if* $g([0,1]) = [0,1]$ *then* $g = e \in \mathrm{BS}(1,2)$. *Thus* $g \sim g([0,1])$ *is a bijection between* $\mathcal{I}$ *and the elements of* $\mathrm{BS}(1,2)$.

*Proof.* Let $[x,y] \in \mathcal{I}$. If $g \in \mathrm{BS}(1,2)$, then by Proposition 4.1 we may express $g$ as $g(x) = 2^n \cdot x + \frac{m}{2^k}$. Thus

$$g([x, x+2^i]) = [2^n \cdot x + \frac{m}{2^k}, 2^n \cdot x + \frac{m}{2^k} + 2^{n+i}].$$

The left endpoint, $2^n \cdot x + \frac{m}{2^k}$, is in $\mathbb{Z}[1/2]$ and the width of the interval is $2^{n+i}$, hence the image of an interval in $\mathcal{I}$ is an interval in $\mathcal{I}$.

The element $a^{-k}b^m a^{k+i}$ is the linear function $g(x) = 2^i \cdot x + \frac{m}{2^k}$, which takes the interval $[0,1]$ to the interval $[\frac{m}{2^k}, \frac{m}{2^k} + 2^i]$, proving that the action of $\mathrm{BS}(1,2)$ on $\mathcal{I}$ is transitive.

Finally, any linear function is determined by the image to two points. If $g([0,1]) = [0,1]$ then either $g$ is the identity, or $g(0) = 1$ and $g(1) = 0$. But this is not possible, since the coefficient of $x$ in $g(x) = 2^n \cdot x + \frac{m}{2^k}$ is positive. $\square$

The group $BS(1,2)$ will show up from time to time in the remainder of this text. In order to give an example of the importance of Baumslag–Solitar groups within geometric group theory, we state a result that we will not prove:

**Theorem 4.3** (See [Me72]). *There is a surjective group homomorphism* $\phi : BS(2,3) \to BS(2,3)$ *(where $\phi(a) = a$ and $\phi(b) = b^2$) that is not an isomorphism.*

### Exercises

(1)  Fill in the outline of the proof of Proposition 4.1.

(2)  Show that $ab = b^n a$ in $BS(1,n)$.

(3)  Show that $\text{Stab}(0) = \langle a \rangle$ for $BS(1,2) \curvearrowright \mathbb{R}$.

(4)  Let $x \in \mathbb{Z}[1/2]$ be any dyadic rational. What is $\text{Stab}(x)$ under the action $BS(1,2) \curvearrowright \mathbb{R}$?

(5)  State and prove the analog of Proposition 4.1 for $BS(1,n)$.

(6)  Prove that the cyclic subgroup of $BS(1,2)$ generated by the element $b$ is not a normal subgroup.

(7)  Prove that the smallest normal subgroup of $BS(1,2)$ that contains $b$ is isomorphic to the dyadic rationals.

(8)  Prove that $BS(1,2)$ is isomorphic to the group of matrices generated by $\begin{pmatrix} 2 & 0 \\ 0 & 1 \end{pmatrix}$ and $\begin{pmatrix} 1 & 1 \\ 0 & 1 \end{pmatrix}$.

# 5

# Words and Dehn's Word Problem

Irgend ein Element der Gruppe ist durch seine Zusammensetzung aus den
Erzeugenden gegeben. Man soll eine Methode angeben, um mit einer endlichen
Anzahl von Schritten zu entscheiden, ob dies Element der Identität gleich ist
oder nicht.

–Max Dehn

## 5.1 Normal Forms

In this chapter we develop connections between the study of infinite
groups and the study of formal languages, a theme that was hinted at in
Section 1.9. We begin by establishing some notation and terminology.

Let $S \subset G$ be a generating set for a group $G$, let "$S$" be a copy of
$S$, thought of as just a set of letters, and let $S^{-1}$ be a set of formal
inverses for the elements of $S$. There is a function from the free monoid
$\{S \cup S^{-1}\}^*$ to $G$:

$$\pi : \{S \cup S^{-1}\}^* \to G.$$

This evaluation map takes each $s \in S$ to the corresponding generator
of $G$, it takes each $s^{-1}$ to the inverse of the corresponding generator of
$G$, and it takes a word $\omega \in \{S \cup S^{-1}\}^*$ to the corresponding product
of generators and their inverses in $G$. Since $S$ is a generating set for
$G$, the map $\pi$ is onto. We will often want to be clear about whether $\omega$
stands for a string of letters or an element in $G$. Unless it is indicated
otherwise, we let $\omega$ be the string of letters and $\pi(\omega)$ the element of $G$.
So, for example, if we let $a = (1,0) \in \mathbb{Z} \oplus \mathbb{Z}$ and $b = (0,1) \in \mathbb{Z} \oplus \mathbb{Z}$,
then the words $\omega_1 = aba^{-1}bbba$ and $\omega_2 = bbabb$ are distinct elements of
$\{a, b, a^{-1}, b^{-1}\}^*$, but $\pi(\omega_1) = \pi(\omega_2) = (1, 4) \in \mathbb{Z} \oplus \mathbb{Z}$.

The fact that elements in a group $G$ have different "spellings" in terms
of the generators is sometimes a cause of confusion. Generally one prefers

105

to have a prescribed method of expressing any given $g \in G$ as a product of the generators and their inverses. A normal form is precisely that. That is, a *normal form* is a function

$$\eta : G \to \{S \cup S^{-1}\}^*$$

such that the composition $\pi \circ \eta : G \to G$ is the identity. In other words, it is a section of the evaluation map $\pi$. Often one thinks of a normal form in terms of the image of $G$ in the free monoid $\{S \cup S^{-1}\}^*$. In this setting, a normal form is simply a subset of $\{S \cup S^{-1}\}^*$ that maps bijectively to $G$ under the evaluation map $\pi$.

**Example 5.1.** The set of freely reduced words in $\{S \cup S^{-1}\}^*$ is a normal form for the free group with basis $S$.

**Example 5.2.** Consider the free abelian group $\mathbb{Z} \oplus \mathbb{Z}$ with generators $\{a, b\}$ as above. Since $a$ and $b$ commute, in any product of $a$'s, $b$'s, and their inverses, one can move the $a$'s and $a^{-1}$'s to the left and the $b$'s and $b^{-1}$'s to the right. This shows that

$$\mathrm{NF}_1 = \{a^i b^j \mid i, j \in \mathbb{Z}\}$$

is a normal form for $\mathbb{Z} \oplus \mathbb{Z}$. There is certainly no claim being made that this is the only normal form. For example, one could push $a$'s and $a^{-1}$'s to the right (and $b$'s and $b^{-1}$'s to the left) and get the resulting normal form: $\mathrm{NF}_2 = \{b^i a^j \mid i, j \in \mathbb{Z}\}$. As yet another example, one could have a preference for alternating $a$'s and $b$'s. Thus instead of writing $a^3 b^4$ one might prefer $abababb = (ab)^3 b$. This leads to the normal form $\mathrm{NF}_3 = \{(ab)^i b^j \mid i, j \in \mathbb{Z}\}$, where for $n \geq 0$ we let $(ab)^n$ be shorthand for the word $\underbrace{ab \cdot ab \cdots ab}_{n \text{ copies of } ab}$ and similarly we let $(ab)^{-n}$ $(n > 0)$ denote the word $(b^{-1} a^{-1})^n$.

Because of the correspondence between words in $\{S \cup S^{-1}\}^*$ and paths in the corresponding Cayley graph (see Section 1.9), a normal form $\eta : G \to \{S \cup S^{-1}\}^*$ can also be thought of as a collection of paths in the Cayley graph $\Gamma_{G,S}$. If $g \in G$ then the word $\eta(g)$ has an associated edge path in $\Gamma_{G,S}$ that begins at the identity and traverses edges as prescribed by $\eta(g)$. For example, consider the normal forms given in Example 5.2. If we think of the generator $a \in \mathbb{Z} \oplus \mathbb{Z}$ as corresponding to the horizontal edges in Figure 5.1, and $b \in \mathbb{Z} \oplus \mathbb{Z}$ as corresponding to the vertical edges, then the paths corresponding to the elements of $\mathrm{NF}_1$ move horizontally, then vertically. Similarly the paths corresponding to

Fig. 5.1. Three paths corresponding to the three normal forms $NF_1$, $NF_2$ and $NF_3$. Each path joins the vertex associated to the identity to the vertex associated to $a^3b^2 = b^2a^3 = (ab)^3b^{-1} \in \mathbb{Z} \oplus \mathbb{Z}$.

words in $NF_2$ move vertically then horizontally. Finally the paths corresponding to the words in $NF_3$ move along the diagonal, and then they either move directly up or down.

**Remark 5.3.** When thought of as a collection of paths in a Cayley graph, a normal form is occasionally referred to as a "combing," although that word is often restricted to sets of normal forms where the associated paths satisfy certain geometric and/or language theoretic restrictions. The terminology is appropriate, at least for "nice" normal forms. Consider for example $G = \mathbb{Z} \oplus \mathbb{Z}_4$ with generators $x = (1,0)$ and $y = (0,1)$. There is a normal form

$$\mathcal{N}(G) = \{x^i y^j \mid i \in \mathbb{Z} \text{ and } 0 \leq j \leq 3\} .$$

If you trace out the paths associated to this normal form, you get something that looks "combed" (Figure 5.2).

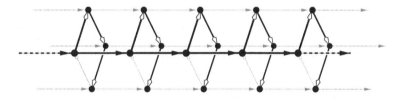

Fig. 5.2. A combing of the Cayley graph of $\mathbb{Z} \oplus \mathbb{Z}_4$. Paths corresponding to the combing are contained in the spanning tree indicated by darker edges.

For a survey of "hairdressing in groups," see [Re98].

In the previous chapter we introduced the Baumslag–Solitar group $BS(1,2)$. In proving various results about the action of this group on $\mathbb{R}$, we developed and then relied on particular expressions for the elements of BS(1,2); we showed that every element of BS(1,2) can be expressed as a word of the form:

$$W = \{a^{-k}b^m a^{k+n} \mid k,m,n \in \mathbb{Z}\}.$$

This set of words is not, however, a normal form. The words $aba^{-1}$ ($k = -1$, $m = 1$ and $n = 0$) and $b^2$ ($k = n = 0$ and $m = 2$) are both in $\mathcal{W}$. But $aba^{-1} = b^2 \in$ BS(1,2). Thus the evaluation map restricted to $\mathcal{W}$, $\pi : \{a,b,a^{-1},b^{-1}\}^* \to$ BS(1,2), is onto BS(1,2), but it is not one-to-one. However, we are not far from a normal form.

**Proposition 5.4.** *The set*

$$\text{NF} = \{a^{-k}b^{2m+1}a^{k+n} \mid k,m,n \in \mathbb{Z}\} \cup \{a^n \mid n \in \mathbb{Z}\}$$

*is a normal form for* BS(1,2).

*Proof.* We know that every $g \in$ BS(1,2) can be expressed as a linear function of the form $g(x) = 2^n x + \frac{\hat{m}}{2^k}$. By reducing to lowest terms we may assume that $\hat{m}$ is either odd or zero. Hence every $g \in$ BS(1,2) has the form $g(x) = 2^n x + \frac{2m+1}{2^k}$ or $g(x) = 2^n x$. The words

$$\{a^{-k}b^{2m+1}a^{k+n} \mid k,m,n \in \mathbb{Z}\}$$

describe the functions $g(x) = 2^n x + \frac{2m+1}{2^k}$ and the words $\{a^n \mid n \in \mathbb{Z}\}$ give $g(x) = 2^n x$, so our set of words maps onto BS(1,2) under $\pi$.

To establish uniqueness, assume to the contrary that

$$a^{-k}b^{2m+1}a^{k+n} = a^{-K}b^{2M+1}a^{K+N}$$

where at least one of the following inequalities holds: $k \neq K$, $m \neq M$ or $n \neq N$. The function on the left-hand side takes

$$[0,1] \mapsto \left[\frac{m}{2^k}, \frac{2m+1}{2^k} + 2^n\right]$$

while the function on the right-hand side takes

$$[0,1] \mapsto \left[\frac{M}{2^K}, \frac{2M+1}{2^K} + 2^N\right].$$

It follows that

$$\frac{2m+1}{2^k} = \frac{2M+1}{2^K} \text{ and } 2^n = 2^N .$$

Thus $m = M$, $k = K$ and $n = N$, which is a contradiction. The same sort of argument shows that no expression of the form $a^{-k}b^{2m+1}a^{k+n}$ yields a function of the form $g(x) = 2^n x$, hence NF is a normal form for BS(1,2). □

## 5.2 Dehn's Word Problem

There are a number of questions in the study of groups that can be thought of as "linguistic" issues. The most famous example of this is Dehn's *word problem*, published in 1912 (see [De12]). Max Dehn asked: *Given a group G and a finite generating set S, is it possible to decide which words in $\{S \cup S^{-1}\}^*$ represent the identity?* In other words, given an arbitrary word $\omega \in \{S \cup S^{-1}\}^*$ is there a way to decide if $\pi(\omega) = e \in G$?

Sometimes it is easy to solve Dehn's word problem. If we generate $\mathbb{Z} \oplus \mathbb{Z}$ by $a = (1,0)$ and $b = (0,1)$, then the words in $\{a, b, a^{-1}, b^{-1}\}$ that represent the identity are precisely the words where the sum of the $a$-exponents and the sum of the $b$-exponents are both zero.

A slightly more complicated case would be the infinite dihedral group $D_\infty$ generated by the reflections $a$ and $b$ (see Chapter 2). Since $a$ and $b$ are their own inverses we do not need to add a set of formal inverses to our generating set; we can simply consider words in $\{a, b\}^*$. Given such a word we can repeatedly apply the following rule:

$D_\infty$-rule: If $\omega \in \{a, b\}^*$ and $\omega$ contains a subword of the form $aa$ or $bb$, then remove this subword to make a new word $\omega'$.

This new word represents the same element of $D_\infty$ as the word we started with, so determining if $\pi(\omega') = e$ is sufficient for determining if the original word represents the identity. There are, potentially, a number of choices that could be made in applying this rule, as is illustrated in Figure 5.2. Any word that results from such a process is equivalent to $w$ – in the sense that its image under the evaluation map is $\pi(w)$ – and the final word one gets is either empty or an alternating sequence of $a$'s and $b$'s. But Proposition 2.1 shows that all alternating sequences of $a$'s and $b$'s yield non-identity elements of $D_\infty$. Thus the word $w$ represents the identity in $D_\infty$ if and only if $w$ reduces to the empty word by repeated application of the $D_\infty$-rule.

There is a variant of Dehn's word problem that asks not only if one can determine when a word represents the identity, but: *given a group G, a finite generating set S, and two words $\omega, \omega' \in \{S \cup S^-\}^*$, can one determine if $\omega$ and $\omega'$ represent the same element of G?* In other words, can you determine if $\pi(\omega) = \pi(\omega')$? This is referred to as the "equality problem."

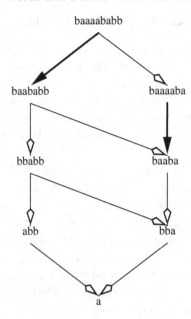

Fig. 5.3. There are a variety of choices that can be made in showing that *baaaababb* = *a* ∈ $D_\infty$. Thicker arrows with solid arrowheads indicate when multiple choices result in the same word.

**Proposition 5.5.** *Given a group* $G$ *and a finite generating set* $S$, *the equality problem can be solved if and only if Dehn's word problem can be solved.*

*Proof.* Since the identity is represented by the empty word, if one can solve the equality problem then one can determine when a given word $\omega$ gives the same element as the empty word, hence one can solve Dehn's word problem.

Conversely, $\omega$ and $\omega'$ represent the same element in $G$ if and only if $\omega' \cdot \omega^{-1}$ represents the identity element. Thus if one can solve Dehn's word problem, one can also solve the equality problem.  □

Dehn's original motivation for formulating and studying the word problem came from issues in algebraic topology. It was eventually discovered that there are finitely generated (and, better yet, finitely presented) groups where the word problem is unsolvable. This was established independently by Novikov and Boone in the mid-1950s. In fact, these techniques establish a quite striking result:

**Theorem 5.6** ([BBN59]). *There are finitely presented groups G such that there is no algorithm that can determine whether a given word in the generators represents:*

1. *the identity;*
2. *an element in the center $Z(G)$;*
3. *an element in the commutator subgroup;*
4. *an element of finite order.*

We will not construct examples of such groups in this book, but they are of great importance in a variety of fields and they have practical applications to areas such as coding theory. For more information, see the survey article [Mi92].

## 5.3 The Word Problem and Cayley Graphs

We commented in Section 1.9 that there is a one-to-one correspondence between words in the generating set of a group and edge paths in the Cayley graph based at the vertex associated to the identity. Given this correspondence there are a number of ways to give a geometric interpretation to questions that seem quite algebraic. Dehn's word problem, for example, can be translated as in the Proposition below.

**Proposition 5.7.** *Given a group G and a finite generating set S, one can solve Dehn's word problem if and only if one can decide which words $\omega$ correspond to closed paths $p_\omega$ in the Cayley graph $\Gamma_{G,S}$.*

**Example 5.8.** The word problem for $D_n$ (with respect to the generating set $\{f, \rho\}$ where $f$ is a reflection and $\rho$ a rotation through an angle of $2\pi/n$) can be solved. To determine if a word $\omega$ represents the identity, simply trace out the corresponding edge path $p_\omega$ in the Cayley graph $\Gamma$. The word $\omega$ represents the identity if and only if the end of the path $p_\omega$ is at the vertex in $\Gamma$ that corresponds to the identity.

This example indicates that the word problem can be solved in any group where the Cayley graph can be constructed. Since our primary focus is on infinite groups, we need an appropriate notion of what it means to say a Cayley graph "can be constructed" if we hope to be able to convert this example to a theorem. A reasonable answer is that one can build arbitrarily large pieces of the Cayley graph, where the idea of "arbitrarily large pieces" is codified by the notion of balls in a Cayley graph.

**Definition 5.9.** Let $\Gamma$ be a connected graph. The *distance* from a vertex $v$ to a vertex $w$ is the minimum number of edges in a combinatorial path joining $v$ to $w$. For example, if $v$ and $w$ are distinct, but form $\text{ENDS}(e)$ for some edge $e$, then the distance from $v$ to $w$ is 1. We denote this distance function by $d_\Gamma(v, w)$.

For a fixed vertex $v \in V(\Gamma)$ and a fixed $n \in \mathbb{N}$, the *sphere of radius $n$ centered at $v$* is the set of vertices

$$\mathcal{S}(v, n) = \{ w \in V(\Gamma) \mid d_\Gamma(v, w) = n \}.$$

The *ball of radius $n$ centered at the vertex $v$* is the subgraph formed as the union of all paths in $\Gamma$ of length $\leq n$ that start at the vertex $v$. We denote this subgraph by $\mathcal{B}(v, n)$. The vertices of $\mathcal{B}(v, n)$ are the union of the vertices in $\mathcal{S}(v, i)$ for $i \leq n$.

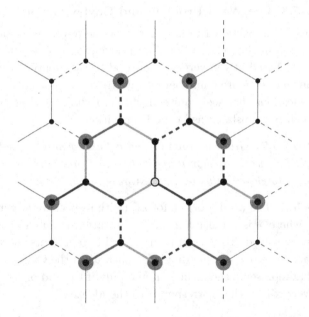

Fig. 5.4. The sphere of radius 3 in the Cayley graph of the reflection group $W_{333}$ (see Chapter 2) consists of all the vertices with a dark halo; the ball of radius 3 is indicated by thicker edges.

In the case of a Cayley graph $\Gamma$ of a group $G$ we abuse notation and write $\mathcal{S}(g, n)$ and $\mathcal{B}(g, n)$ to denote the sphere and ball of radius $n$ centered at the vertex associated to $g \in G$. (An example of a ball and sphere in a Cayley graph is shown in Figure 5.4.)

A Cayley graph $\Gamma$ is *constructible* if given $n \in \mathbb{N}$ one can construct $\mathcal{B}(e, n)$ in a finite amount of time.

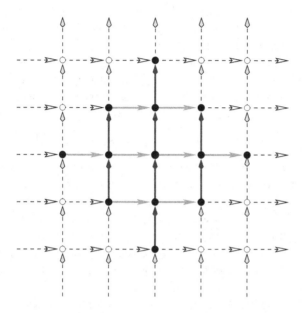

Fig. 5.5. The ball of radius 2 in the Cayley graph of $\mathbb{Z} \oplus \mathbb{Z}$ consists of all the dark vertices and solid arrows.

**Theorem 5.10.** *Let $G$ be a group with finite generating set $S$. Then the word problem is solvable if and only if the Cayley graph $\Gamma_{G,S}$ is constructible.*

*Proof.* ($\Leftarrow$) Let $\omega \in \{S \cup S^{-1}\}^*$ and let $n = |\omega|$. Construct the subgraph $\mathcal{B}(e, n) \subset \Gamma_{G,S}$, and trace out the path $p_\omega$ inside this ball. The terminal vertex of this path is $v_e$ if and only if $\pi(\omega) = e \in G$. Thus one can solve the word problem.

($\Rightarrow$) The proof that $B(e, n)$ is constructible is by induction on $n$. The base case is immediate: $B(e, 0)$ consists of a single vertex corresponding to the identity.

Assume that $\mathcal{B}(e, n)$ has been constructed, so that our goal is to construct $\mathcal{B}(e, n + 1)$. Every vertex in $\Gamma_{G,S}$ is incident with exactly $k$ edges corresponding to the generators $S \cup S^{-1}$. If $v \in B(e, n - 1)$, then $v$ is already incident with $k$ edges labelled by $S \cup S^{-1}$. If $d_\Gamma(v_e, v) = n$, let $\text{LABELS}_n(v)$ be the subset of $S \cup S^{-1}$ corresponding to edges that are attached to $v$ in $\mathcal{B}(e,n)$. In forming $\mathcal{B}(e,n+1)$ we need to add an (oriented) edge to $v$, labelled $s$, for each $s \in \{S \cup S^{-1}\} \setminus \text{LABELS}_n(v)$. The trick is to discover what the other vertex is that bounds this edge. Notice that none of the missing edges can join $v_i$ to a vertex in $\mathcal{B}(e,n-1)$. If one did, then the edge in question would have been present in $\mathcal{B}(e,n)$, and so its label would be in $\text{LABELS}_n(v_i)$.

For convenience, list the elements in the sphere of radius $n$ in some fixed order: $\mathcal{S}(e,n) = \{v_1, v_2, \ldots, v_l\}$. To each $v_i$ associate a word $\omega_i$ whose corresponding path in $\mathcal{B}(e,n)$ joins the vertex associated to $e$ to $v_i$. Starting with vertex $v_1$, list the generators in

$$\{S \cup S^{-1}\} \setminus \text{LABELS}_n(v_1) = \{s_1, s_2, \ldots, s_m\}.$$

The edge of $\Gamma_{G,S}$ that is attached to $v_1$ and is labelled $s_i$ cannot join $v_1$ to a vertex in $\mathcal{B}(e,n-1)$, so it must either join $v_1$ to some other $v_j \in \mathcal{S}(e,n)$, or join $v_1$ to a vertex in $\mathcal{S}(e,n+1)$. This edge joins $v_1$ to $v_j \in \mathcal{S}(e,n)$ if and only if the word $\omega_1 s_i (\omega_j)^{-1}$ describes a circuit in $\Gamma_{G,S}$, that is, if and only if $\pi\left(\omega_1 s_i (\omega_j)^{-1}\right) = e \in G$. Since we can solve the word problem, we can check this condition. If $\pi\left(\omega_1 s_i (\omega_j)^{-1}\right) \neq e \in G$ for each $j$, then the edge labelled $s_i$ joins $v_1$ to a new vertex, not seen in $\mathcal{B}(e,n)$.

Let $i > 1$ and let $s_j$ be a missing label for the vertex $v_i$, that is, $s_j \in \{S \cup S^{-1}\} \setminus \text{LABELS}_n(v_i)$. The edge labelled $s_j$, which is attached to $v_i$ in $\Gamma_{G,S}$, joins $v_i$ to one of three types of vertices:

1. another $v_k \in \mathcal{S}(e,n)$;
2. a vertex in $\mathcal{S}(e,n+1)$ that is joined to $v_m \in \mathcal{S}(e,n)$, where $m < i$; or
3. a vertex in $\mathcal{S}(e,n+1)$ that is not attached to any of the previous vertices $\{v_1, \ldots, v_{i-1}\}$.

(The second case does not occur when $i = 1$.) Case 1 occurs if and only if $\pi\left(\omega_i s_j (\omega_m)^{-1}\right) = e \in G$, as before. Similarly, case 2 occurs if and only if $\pi\left(\omega_i s_j (\omega_m s_n)^{-1}\right) = e \in G$ for some $s_n \in \{S \cup S^{-1}\}$, and this can be checked since we can solve the word problem. If none of the appropriate words represent the identity, then we know we are in Case 3, and a new vertex needs to be added in forming $\mathcal{B}(e,n+1)$. As any vertex in $\mathcal{B}(e,n+1)$ is joined to a vertex in $\mathcal{B}(e,n)$, this process will terminate in a finite number of steps. $\square$

**Example 5.11.** In Section 3.5 we explored the free product $G = \mathbb{Z}_3 * \mathbb{Z}_4$. This group is generated by an element $a$ of order 4 (corresponding to

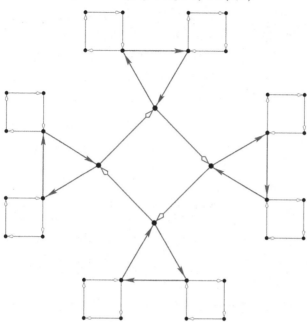

Fig. 5.6. Part of the Cayley graph of $\mathbb{Z}_3 * \mathbb{Z}_4$ with respect to the generating set $\{a, b\}$.

the $\mathbb{Z}_4$) and an element $b$ of order 3 (corresponding to the $\mathbb{Z}_3$). We can construct a normal form for $G$ by remembering that the elements of $G$ can be expressed as words that alternate between non-identity elements of $\mathbb{Z}_3$ and $\mathbb{Z}_4$. These non-identity elements can be uniquely expressed by the words $\{a, a^2, a^3, b, b^2\}$. Knowing this, and applying the method described in the proof of Theorem 5.10, one can verify that the Cayley graph of $G$ with respect to the generating set $\{a, b\}$ looks like a "tree of squares and triangles" as in Figure 5.6.

**Remark 5.12.** In discussing the distance between vertices in a Cayley graph, and the notion of spheres and balls of a given radius, we have touched on the geometric side of group theory. This topic is explored in Chapter 9.

## 5.4 The Cayley Graph of BS(1,2)

It is easy to solve the word problem for BS(1, 2). Given any word $\omega \in \{a, b, a^{-1}, b^{-1}\}^*$ one can produce the associated group element by composing the appropriate linear functions. In the end one has $g(x) = x$ if and only if $\pi(\omega) = e$. For example, consider the word

$$\omega = a^2 b a^{-2} b a^2 b^{-1} a^{-2} b^{-1},$$

and let $g = \pi(\omega)$. Then

$$g : x \xrightarrow{b^{-1}} x - 1 \xrightarrow{a^{-2}} \frac{1}{4}x - \frac{1}{4} \xrightarrow{b^{-1}} \frac{1}{4}x - \frac{5}{4} \xrightarrow{a^2} x - 5$$

$$\xrightarrow{b} x - 4 \xrightarrow{a^{-2}} \frac{1}{4}x - 1 \xrightarrow{b} \frac{1}{4}x \xrightarrow{a^2} x$$

so $g = \pi(\omega) = e \in \mathrm{BS}(1,2)$.

Since we can solve the word problem, Theorem 5.10 implies that we should be able to build the Cayley graph of $\mathrm{BS}(1,2)$ relative to the generators $\{a,b\}$. Because $ab = b^2 a$, or equivalently $aba^{-1} = b^2$, the Cayley graph must contain a 5-cycle corresponding to this relation. This 5-cycle is informally referred to as a *brick* in the Cayley graph. We can

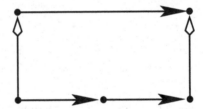

Fig. 5.7. The "brick" corresponding to $ab = b^2 a$, where horizontal arrows correspond to the generator $b$, and vertical arrows to $a$.

begin to construct the Cayley graph by stacking these bricks together. The resulting grid of bricks can be continued indefinitely, forming what is referred to as a *sheet* in the Cayley graph of $\mathrm{BS}(1,2)$, as is shown in Figure 5.8.

The sheet shown in Figure 5.8 is not the complete Cayley graph of $\mathrm{BS}(1,2)$ with respect to $\{a,b\}$. One can see that there must be some edges missing, as some of the vertices in this grid have only three edges incident with them, not four. For example, if you assume the identity is the vertex in the lower left corner of Figure 5.8, you cannot trace out the final $a$ in the word $aba$. Because we know $aba$ is the unique representation of an element of $\mathrm{BS}(1,2)$ (in our normal form), and because every vertex in this Cayley graph must be incident to four edges, there must be a missing vertex and edge. One standard way to depict this is to have the missing edges come out from the sheet, as in Figure 5.9.

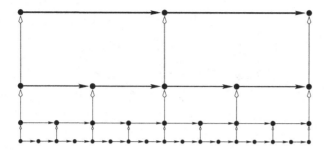

Fig. 5.8. A "sheet" in the Cayley graph of BS(1,2).

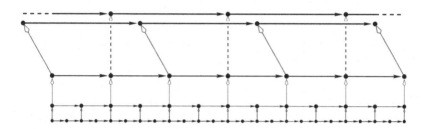

Fig. 5.9. More of the Cayley graph of BS(1,2).

Given any vertex in the Cayley graph, one can form a combinatorial line by right multiplying by positive and negative powers of $b$. Call such a line a *b-line* in the Cayley graph of BS(1,2). Each such line will have one set of edges labelled $a$ coming into it and two sets of edges labelled $a$ pointing away from it, corresponding to whether the exponent of $b$ is odd or even. Figure 5.9 illustrates the fact that two distinct sheets come together at each $b$-line. These sheets merge below the $b$-line (the portion corresponding to $a$-edges which point into the $b$-line) but are distinct above the $b$-line. This is why, informally speaking, a side view of the Cayley graph of BS(1,2) looks like a trivalent tree (Figure 5.10).

At this stage the reader is well-poised to think about our characterization of BS(1,2) given in Corollary 4.2. There it is shown that there is a one-to-one correspondence between elements of BS(1,2) and the images of the unit interval $[0,1]$ under the action of BS(1,2). Let $\phi : \Gamma \to \mathbb{R}$ be the map induced by sending the vertex associated to $b^n$ to $n \in \mathbb{R}$ and by requiring that the image of any edge labelled $a$ is a single point, as indicated in Figure 5.11. We leave the proof of the following insight to the exercises.

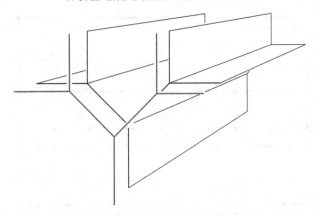

Fig. 5.10. The Cayley graph of BS(1, 2) projects onto a trivalent tree.

**Proposition 5.13.** *Let $g$ be an element of* BS(1, 2). *Let $e_g$ be the edge in the Cayley graph $\Gamma$ that is labelled $b$ and has the vertex associated to $g$ as its initial vertex. Then $\phi(e_g) = g([0, 1])$.*

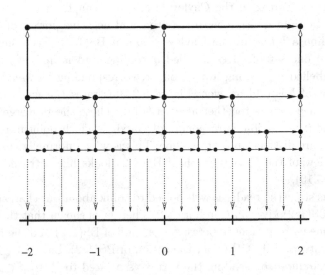

Fig. 5.11. Projecting the Cayley graph of BS(1, 2) onto $\mathbb{R}$.

## Exercises

(1) Three different normal forms for $\mathbb{Z} \oplus \mathbb{Z}$ are given in Example 5.2. Draw the Cayley graph of $\mathbb{Z} \oplus \mathbb{Z}$ and the paths corresponding to the words in $\mathrm{NF}_1, \mathrm{NF}_2$ and $\mathrm{NF}_3$ for the elements:

    a. $(3,5)$
    b. $(-2,-1)$
    c. $(1,6)$

(2) Find a normal form for $\mathbb{F}_2 \oplus \mathbb{F}_2$.

(3) Use colored pencils to draw spheres of radius $n$, for $n = 1, 2$ and $3$, in the Cayley graph of:

    a. the cyclic group $\mathbb{Z}_6$ (with respect to a cyclic generator).
    b. the symmetric group $\mathrm{SYM}_4$ (with respect to the adjacent transpositions $\{(12), (23), (34)\}$).
    c. the reflection group $W_{333}$ from Chapter 2 (with respect to the three reflections in the sides of the equilateral triangle).
    d. $\mathrm{BS}(1,2)$ (with respect to $\{a,b\}$).

(4) Prove that if you can solve the word problem for groups $A$ and $B$ then you can solve it for $G = A \oplus B$.

(5) Prove that if you can solve the word problem for groups $A$ and $B$ then you can solve it for $G = A * B$.

(6) Solve the word problem for the reflection group $W_{333}$ from Chapter 2.

(7) Show that $\pi(a^3 b^2 a^{-3} b a^3 b^{-2} a^{-3} b^{-1}) = e \in \mathrm{BS}(1,2)$.

(8) Prove Proposition 5.13.

(9) Construct the Cayley graph of $\mathrm{BS}(1,3)$.

(10) In addition to the word problem, Max Dehn also introduced the following *conjugacy problem* in his 1912 paper.

    *Let $S$ be a finite generating set for a group $G$. Let $\omega$ and $\omega'$ be words in $\{S \cup S^{-1}\}^*$. Can one determine if $\pi(\omega)$ and $\pi(\omega')$ are conjugate?*

    Solve the conjugacy problem in the infinite dihedral group $D_\infty$.

(11) Show that you can solve the conjugacy problem (see Exercise 10) in $A * B$ if you can solve it in $A$ and $B$.

# 6

# A Finitely Generated, Infinite Torsion Group

In 1902, William Burnside asked the following natural question:

*Does every finitely generated infinite group contain an element of infinite order?*

Today this is known as the *Weak Burnside Problem* and, while it appears to be an elementary question, it took over sixty years to resolve. On the face of it, one would probably believe the answer is "yes." The infinite groups we have encountered so far all have elements of infinite order. This is true for $\mathbb{Z}^n$, $\mathbb{F}_n$, $A * B$ where $A$ and $B$ are non-trivial groups, and BS(1, 2). However, in this chapter we construct an example that shows that the answer is actually "no."

A group is *torsion-free* if no non-identity element has finite order; examples include the free groups and BS(1, 2). A group is a *torsion group* if every element has finite order; the only examples you are likely to have encountered are finite groups. These two conditions are at opposite ends of a spectrum. There are a number of groups that are neither torsion-free nor torsion, such as $\mathbb{Z}_2 * \mathbb{Z}_3$. In this terminology, Burnside was asking if there are finitely generated, infinite torsion groups.

There are easy ways to construct infinite torsion groups: one can form a countable direct sum of finite cyclic groups, such as

$$\mathbb{Z}_2 \oplus \mathbb{Z}_2 \oplus \mathbb{Z}_2 \oplus \cdots$$

and the factor group $\mathbb{Q}/\mathbb{Z}$ is also an infinite torsion group. But in neither of these examples is the group finitely generated. (Establishing these claims is your Exercise 1.) To construct our example of a finitely generated, infinite torsion group we use a variation on the theme of Chapter 3

120

(groups acting on trees), and describe our group via symmetries of a rooted tree.

**Definition 6.1.** A *rooted* tree consists of a tree $T$ and a selected vertex $v$, called the *root*. In contrast to common experience, rooted trees are generally drawn with the root at the top, with edges and other vertices descending downward (as in Figure 6.1). Using this convention one can define the *children* of a vertex $v \in T$ to be those vertices that are immediately below $v$. They can also be defined as the vertices that are adjacent to $v$ which are not on the unique, reduced edge path joining $v$ to the root. Similarly, the *descendants* of $v$ are those vertices that are joined to $v$ by a descending path, or more formally, a vertex $w$ is a descendant of $v$ if the only point of intersection between the unique reduced edge path joining $w$ to $v$ and the unique reduced edge path joining $v$ to the root, is the vertex $v$.

A rooted tree where each vertex which is not a leaf has $n$ children is referred to as a *rooted n-ary tree*.

The group $\mathrm{SYM}(T)$, for a rooted tree $T$, consists of all symmetries of $T$ that fix the root.

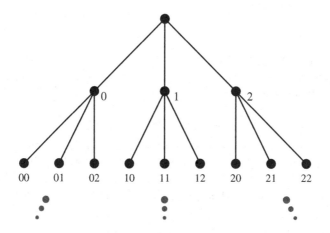

Fig. 6.1. The rooted, infinite ternary tree $T$.

We let $T$ denote the rooted ternary tree for the remainder of this chapter. We label the vertices of this tree using finite sequences $n_1 n_2 \cdots n_k$ where each $n_i \in \mathbb{Z}_3$. The root corresponds to the empty sequence; the vertex associated to $n_1 \cdots n_k n_{k+1}$ is a child of the vertex associated to $n_1 \cdots n_k$; and the vertex associated to $n_1 \cdots n_k n_{k+1} \cdots n_l$ is a descendant of the vertex associated to $n_1 \cdots n_k$. Denote the vertex associated to the finite sequence $n_1 \cdots n_k$ by $v_{n_1 \cdots n_k}$, and define the *vertices at level $k$* in $\mathcal{T}$ to be those vertices whose labels are sequences of length $k$. As an example, $\{v_0, v_1, v_2\}$ are the vertices at level 1. The first two levels of $\mathcal{T}$, with the vertices labelled, are illustrated in Figure 6.1. Notice that, because the root is fixed, any symmetry of the rooted ternary tree $\mathcal{T}$ must preserve the levels.

Let $\sigma \in \mathrm{SYM}(\mathcal{T})$ be the symmetry of $\mathcal{T}$ described by

$$\sigma(v_{n_1 n_2 \cdots n_k}) = v_{(n_1+1)n_2 \cdots n_k}$$

where the addition in the first digit of the subscript is done in $\mathbb{Z}_3$. Thus $\sigma(v_0) = v_1$, $\sigma(v_1) = v_2$ and $\sigma(v_2) = v_0$. Notice that, following the drawing conventions of Figure 6.1, the subtree hanging below $v_0$ is taken by $\sigma$ to the subtree hanging below $v_1$ without any twisting or reflecting. One can think of $\sigma$ is something like a shuffle of a deck of cards.

Any vertex $v_{n_1 \cdots n_k} \in \mathcal{T}$ can serve as the root of the subtree of $\mathcal{T}$ induced by $v_{n_1 \cdots n_k}$ and its descendants. Denote this subtree as $\mathcal{T}_{n_1 \cdots n_k}$. One can perform a version of the shuffle $\sigma$ on $\mathcal{T}_{n_1 \cdots n_k}$. This symmetry will fix any vertex whose label does not begin with the sequence $n_1 \cdots n_k$, and it will take

$$v_{n_1 \cdots n_k n_{k+1} n_{k+2} \cdots n_l} \mapsto v_{n_1 \cdots n_k (n_{k+1}+1) n_{k+2} \cdots n_l}.$$

Denote this symmetry by $\sigma_{n_1 \cdots n_k}$.

We can now define a more complicated symmetry of $\mathcal{T}$ as an infinite product of shuffles:

$$\omega = \sigma_0 \sigma_1^{-1} \sigma_{20} \sigma_{21}^{-1} \sigma_{220} \sigma_{221}^{-1} \cdots$$

The symmetry $\sigma_0$ shuffles the subtrees below $v_0$; the symmetry $\sigma_1^{-1}$ shuffles the subtrees below $v_1$ (in the opposite direction). The vertices $v_2$ and its children $v_{20}, v_{21}$, and $v_{22}$ are all fixed by $\omega$. However, the subtrees below $v_{20}$ are shuffled, as are the subtrees below $v_{21}$ (in the opposite direction). In general, $\omega$ will fix any vertex whose label is $222 \cdots 2i$, and it will shuffle the subtrees below $v_{222 \cdots 20}$ and $v_{222 \cdots 21}$ (in the opposite direction). We refer to this symmetry as a *waterfall* (Figure 6.2).

**Theorem 6.2.** *Let $\mathcal{U}$ be the subgroup of $\mathrm{SYM}(\mathcal{T})$ generated by $\sigma$ and $\omega$. Then $\mathcal{U}$ is a finitely generated, infinite group where every $g \in G$ has finite order. Further, the order of any element is a power of 3.*

Before turning to the proof of this result, we set up a bit of useful notation. Just as $\sigma_{n_1 \cdots n_k}$ denotes a shuffle based at the vertex $v_{n_1 \cdots n_k}$,

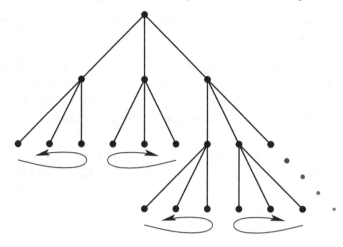

Fig. 6.2. The "waterfall" symmetry $\omega$.

let $\omega_{n_1 \cdots n_k}$ denote a waterfall based at $v_{n_1 \cdots n_k}$. More formally, $\omega_{n_1 \cdots n_k}$ can be written as an infinite product of shuffles, where one adds $n_1 \cdots n_k$ as a prefix to all the indices used in describing $\omega$. For example, $\omega_{21}$ would be

$$\omega_{21} = \sigma_{210}\sigma_{211}^{-1}\sigma_{2120}\sigma_{2121}^{-1}\sigma_{21220}\sigma_{21221}^{-1} \cdots$$

**Definition 6.3.** There are a number of subgroups of $\mathrm{SYM}(\mathcal{T})$ that are isomorphic to $\mathcal{U}$. Namely, if $n_1 n_2 \cdots n_k$ is a label, let $\mathcal{U}_{n_1 n_2 \cdots n_k}$ be the subgroup generated by $\sigma_{n_1 n_2 \cdots n_k}$ and $\omega_{n_1 n_2 \cdots n_k}$. This subgroup of $\mathrm{SYM}(\mathcal{T})$ acts on the subtree $\mathcal{T}_{n_1 n_2 \cdots n_k}$ exactly like $\mathcal{U}$ acts on $\mathcal{T}$.

Using this notation we notice the following formulas, which the reader can verify:

$$\omega = \sigma_0 \sigma_1^{-1} \omega_2,$$
$$\sigma\omega\sigma^{-1} = \omega_0 \sigma_1 \sigma_2^{-1}, \tag{6.1}$$
$$\sigma^2 \omega \sigma^{-2} = \sigma_0^{-1} \omega_1 \sigma_2.$$

The following is another set of formulas that are useful in the proofs of Lemmas 6.4 and 6.5:

$$\omega\sigma = \sigma(\sigma^2 \omega \sigma^{-2}),$$
$$(\sigma\omega\sigma^{-1})\sigma = \sigma\omega, \tag{6.2}$$
$$(\sigma^2 \omega \sigma^{-2})\sigma = \sigma(\sigma\omega\sigma^{-1}).$$

**Lemma 6.4.** *Each $g \in \mathcal{U}$ can be expressed as $g = \sigma^k x_1 \cdots x_n$ where each $x_i \in \{\omega, \sigma\omega\sigma^{-1}, \sigma^2\omega\sigma^{-2}\}$.*

*Proof.* Since $\mathcal{U}$ is the group generated by $\sigma$ and $\omega$, every $g \in \mathcal{U}$ can be expressed as a word in $\{\sigma, \sigma^{-1}, \omega, \omega^{-1}\}^*$. Further, since $\sigma^2 = \sigma^{-1}$ and $\omega^2 = \omega^{-1}$, every $g \in \mathcal{U}$ can be expressed as a word in $\{\sigma, \omega\}^*$. Our goal is to convert such an expression into an expression of the form indicated above.

Let $g = \sigma^j y_1 y_2 \cdots y_n$ where each $y_i \in \{\sigma, \omega\}$ and $y_1 = \omega$. Let $y_k$ be the leftmost $\sigma$ in the suffix $y_1 y_2 \cdots y_n$. Then $y_{k-1} = \omega$. By the first equation given in Equations 6.2,

$$y_{k-1}y_k = \omega\sigma = \sigma(\sigma^2\omega\sigma^{-2}).$$

Thus this $\sigma$ can be moved to the left, converting $\omega$'s to $(\sigma^2\omega\sigma^{-2})$'s as it goes. The next $\sigma$ in $y_1 y_2 \cdots y_n$ can also be moved to the left, converting $\omega$'s to $(\sigma^2\omega\sigma^{-2})$'s, and $(\sigma^2\omega\sigma^{-2})$'s to $(\sigma\omega\sigma^{-1})$'s. Continuing in this manner, we convert $\sigma^j y_1 y_2 \cdots y_n$ into the specified form.  □

We will need to refer to the process described above in proofs that occur below. To facilitate this, we define an operation $\Sigma$ on words $x_1 \cdots x_n \in \{\omega, \sigma\omega\sigma^{-1}, \sigma^2\omega\sigma^{-2}\}$ by first defining $\Sigma$ on the three basic elements

$$\Sigma(\omega) = (\sigma^2\omega\sigma^{-2}),$$
$$\Sigma(\sigma\omega\sigma^{-1}) = \omega,$$
$$\Sigma(\sigma^2\omega\sigma^{-2}) = (\sigma\omega\sigma^{-1}),$$

and then we extend $\Sigma$ to words using $\Sigma(x_1 \cdots x_n) = \Sigma(x_1) \cdots \Sigma(x_n)$. As was exploited in the argument above, Equations 6.2 simply state that $x_i\sigma = \sigma\Sigma(x_i)$ for each $x_i \in \{\omega, \sigma\omega\sigma^{-1}, \sigma^2\omega\sigma^{-2}\}$. Repeated application of this fact shows that $x_1 \cdots x_n\sigma = \sigma\Sigma(x_1 \cdots x_n)$ for any word $x_1 \cdots x_n \in \{\omega, \sigma\omega\sigma^{-1}, \sigma^2\omega\sigma^{-2}\}^*$.

As an example, if we use this notation and apply the process described in the proof of Lemma 6.4 to the word $\omega\omega\sigma\omega\sigma\omega\sigma\omega\omega\sigma$ we get

$$
\begin{aligned}
\omega\omega\sigma\omega\sigma\omega\sigma\omega\omega\sigma \;&\rightarrow\; \sigma\Sigma(\omega)\Sigma(\omega)\omega\sigma\omega\sigma\omega\omega\sigma \\
&\rightarrow\; \sigma^2\Sigma^2(\omega)\Sigma^2(\omega)\Sigma(\omega)\omega\sigma\omega\omega\sigma \\
&\rightarrow\; \sigma^3\Sigma^3(\omega)\Sigma^3(\omega)\Sigma^2(\omega)\Sigma(\omega)\omega\omega\sigma \\
&\rightarrow\; \sigma^4\Sigma^4(\omega)\Sigma^4(\omega)\Sigma^3(\omega)\Sigma^2(\omega)\Sigma(\omega)\Sigma(\omega) \\
&=\; \sigma(\sigma^2\omega\sigma^{-2})^2(\omega)(\sigma\omega\sigma^{-1})(\sigma^2\omega\sigma^{-2})^2
\end{aligned}
$$

where the final equality comes from applying $\Sigma$ and using the fact that the group element $\sigma$ and the operation $\Sigma$ both have order 3.

**Lemma 6.5.** *There is a homomorphism* $\phi : \mathcal{U} \to \mathbb{Z}_3$ *whose kernel* $K$ *is the subgroup generated by the elements* $\{\omega, \sigma\omega\sigma^{-1}, \sigma^2\omega\sigma^{-2}\}$.

*Proof.* Since $\sigma$ and $\omega$ send the vertices of level $k$ to the vertices of level $k$ (for any $k \geq 0$), the group $\mathcal{U}$ sends vertices at level $k$ to vertices at level $k$. Thus for a fixed $k$ there is a map from $\mathcal{U}$ to $\mathrm{SYM}_{3^k}$. In this proof we focus on the action of $\mathcal{U}$ on the three vertices $\{v_0, v_1, v_2\}$. The symmetry $\sigma$ induces the permutation $(123)$ (on the indices) while $\omega$ induces the trivial permutation. Thus the only permutations that occur in the image of $\phi$ are

$$\text{Image of } \phi = \{e, (123), (132)\} = A_3;$$

so $\phi : \mathcal{U} \to A_3 \approx \mathbb{Z}_3$. Since the subgroup generated by $\sigma$ maps onto the quotient, $\phi$ is surjective.

Because $\omega$ is in the kernel of $\phi$, and kernels are normal subgroups, the set $\{\omega, \sigma\omega\sigma^{-1}, \sigma^2\omega\sigma^{-2}\}$ is contained in $K$. The reader can check that the subgroup $H$, generated by the set $\{\omega, \sigma\omega\sigma^{-1}, \sigma^2\omega\sigma^{-2}\}$, is a normal subgroup of $\mathcal{U}$ that is contained in $K$.

If $g \in \mathcal{U}$ then by Lemma 6.4 we can express $g$ as $\sigma^k x_1 \cdots x_n$ where each $x_i \in \{\omega, \sigma\omega\sigma^{-1}, \sigma^2\omega\sigma^{-2}\}$. Thus every $g \in \mathcal{U}$ can be expressed as $g = \sigma^k h$ for $k \in \{0, 1, 2\}$ and $h \in H$. Thus

$$\mathcal{U}/H = \{H, \sigma H, \sigma^2 H\} \approx \mathbb{Z}_3.$$

Therefore $H$ is a normal subgroup of $\mathcal{U}$, contained in $K$, with the same quotient. So the kernel $K$ is actually the subgroup we (temporarily) referred to as $H$, which is generated by $\{\omega, \sigma\omega\sigma^{-1}, \sigma^2\omega\sigma^{-2}\}$. $\square$

As $\mathcal{U}$ is defined to be the group generated by $\sigma$ and $\omega$, the claim that $\mathcal{U}$ is finitely generated is immediate. The claim that $\mathcal{U}$ is infinite follows from Lemma 6.6 below, since in a finite group no proper subgroup can map onto the larger group.

**Lemma 6.6.** *The kernel* $K$ *surjects onto* $\mathcal{U}$.

*Proof.* The subgroup $K$ is the collection of elements in $\mathcal{U}$ that fix $v_0$, $v_1$ and $v_2$. Each of the generators of $K$, $\{\omega, \sigma\omega\sigma^{-1}, \sigma^2\omega\sigma^{-2}\}$, can be expressed as a product of elements of the form $\sigma_i$, $\sigma_i^{-1}$ and $\omega_i$ $(i \in \mathbb{Z}_3)$, as in Equations 6.1. Since the three subtrees $\mathcal{T}_0$, $\mathcal{T}_1$ and $\mathcal{T}_2$ do not intersect, each generator of $K$ can be viewed as an element of $\mathcal{U}_0 \times \mathcal{U}_1 \times \mathcal{U}_2$, where each $\mathcal{U}_i$ is as in Definition 6.3. It follows that $K < \mathcal{U}_0 \times \mathcal{U}_1 \times \mathcal{U}_2$. We may then project $K$ onto any one coordinate, giving a map $\pi_i : K \to \mathcal{U}_i$. For example, under $\pi_0$ we have

$$
\begin{aligned}
\omega &\mapsto \sigma_0, \\
\sigma\omega\sigma^{-1} &\mapsto \omega_0, \\
\sigma^2\omega\sigma^{-2} &\mapsto \sigma_0^{-1}.
\end{aligned}
$$

The image of $\pi_0$ is then the subgroup generated by $\sigma_0$ and $\omega_0$, hence the map $\pi_0 : H \to \mathcal{U}_0$ is onto, and the result follows since $\mathcal{U}_0 \approx \mathcal{U}$. $\qquad\square$

We now turn to the proof of most difficult claim, that every element has order a power of 3.

We may define a "length" for each $g \in \mathcal{U}$ by finding an expression $g = \sigma^k x_1 \cdots x_n$, where $x_1 \cdots x_n \in \{\omega, \sigma\omega\sigma^{-1}, \sigma^2\omega\sigma^{-2}\}^*$ and $n$ is as small as possible. We then define

$$
\ell(g) = \left\{ \begin{array}{ll} n & \text{if } k = 0 \\ n+1 & \text{if } k \neq 0. \end{array} \right.
$$

**Lemma 6.7.** *Every $g \in \mathcal{U}$ has order a power of 3.*

*Proof.* The proof is by induction on $\ell(g)$. The bases case, where $\ell(g) = 1$, consists of showing that all of the elements $\sigma, \sigma^2, \omega, \sigma\omega\sigma^{-1}$ and $\sigma^2\omega\sigma^{-2}$, have order 3.

Assume for the induction step that $\ell(g) = n+1$. There are three cases to consider:

Case 1: $g = \sigma x_1 \cdots x_n$.
Let $n_0$ be the number of occurrences of $\omega$ among the $x_i$, let $n_1$ be the number of occurrences of $(\sigma\omega\sigma^{-1})$ among the $x_i$, and let $n_2$ be the number of occurrences of $(\sigma^2\omega\sigma^{-2})$. Thus $n = n_0 + n_1 + n_2$.

Since we have assumed that $g = \sigma x_1 \cdots x_n$ it follows that

$$
\begin{aligned}
g^3 &= \sigma x_1 \cdots x_n \cdot \sigma x_1 \cdots x_n \cdot \sigma x_1 \cdots x_n \\
&= \Sigma^2(x_1 \cdots x_n)\Sigma(x_1 \cdots x_n)(x_1 \cdots x_n) \in K,
\end{aligned}
$$

where we are using the $\Sigma$ notation and the "sliding $\sigma$ left" technique from the proof of Lemma 6.4. In the resulting expression $\omega$ occurs $n_0$ times in $x_1 \cdots x_n$, $n_1$ times in $\Sigma(x_1 \cdots x_n)$ and $n_2$ times in $\Sigma^2(x_1 \cdots x_n)$. Thus in total $\omega$ occurs $n = n_0 + n_1 + n_2$ times. Similarly $(\sigma\omega\sigma^{-1})$ occurs $n$ times and $(\sigma^2\omega\sigma^{-2})$ occurs $n$ times.

Viewing $K$ as a subgroup of $\mathcal{U}_0 \times \mathcal{U}_1 \times \mathcal{U}_2$, as in the proof of Lemma 6.6, we see that $g^3 = (g_0, g_1, g_2)$, where each $g_i \in \mathcal{U}_i$. Applying the expressions in Equations 6.1 we see that each $g_i$ can be expressed as a word in $\{\omega_i, \sigma_i, \sigma_i^{-1}\}^*$ of length $3n$, with exactly $n$ occurrences of each element $\omega_i$, $\sigma_i$ and $\sigma_i^{-1}$. Using Equations 6.2 we see that the $\sigma_i$ and $\sigma_i^{-1}$ can be moved to the left and cancelled, resulting in a word of length $n$ using the alphabet $\{\omega_i, \sigma_i \omega_i \sigma_i^{-1}, \sigma_i^2 \omega_i \sigma_i^{-2}\}$. By induction, the order of $g_i$ is then a power of 3, making the order of $g^3$ a power of 3, hence the order of $g$ is a power of 3.

Case 2: $g = \sigma^2 x_1 \cdots x_n$.

This is quite similar to Case 1, and is left as an exercise.

Case 3: $g = x_1 \cdots x_n x_{n+1}$.

Here, borrowing notation from Case 1, we have $n_0 + n_1 + n_2 = n + 1$. We may assume that two of $n_0, n_1$, and $n_2$ are non-zero, as otherwise $g$ is a power of $\omega, \sigma\omega\sigma^{-1}$ or $\sigma^2\omega\sigma^{-2}$ in which case $g$ would have order 1 or 3. Thus we may assume that each $n_i$ is less than or equal to $n$.

Since the $x_i \in K$ we know $g \in K$ and therefore

$$g = (g_0, g_1, g_2) \in \mathcal{U}_0 \times \mathcal{U}_1 \times \mathcal{U}_2.$$

Using Equations 6.1 we see that each $g_i$ can be expressed as a word of length $n + 1$ in $\{\sigma_i, \sigma_i^{-1}, \omega_i\}^*$. In fact, we know exactly how many times each of the elements of this alphabet shows up. For example, $\omega_0$ occurs $n_1$ times in the expression for $g_0$, $\omega_1$ occurs $n_2$ times in the expression for $g_1$, and $\omega_2$ occurs $n_0$ times in the expression for $g_2$. We can replace each $\sigma_i^{-1}$ that occurs in the expression for $g_i$ by $\sigma_i^2$. Then, applying the "sliding $\sigma$ to the left" technique of Lemma 6.4, we can express each $g_i$ as

$$g_i = \sigma_i^j x_1 x_2 \cdots x_{m_i}$$

where $x_1 x_2 \cdots x_{m_i} \in \{\omega_i, \sigma_i \omega_i \sigma_i^{-1}, \sigma_i^2 \omega_i \sigma_i^{-2}\}^*$ and $m_i = n_{i+1}$. If $j = 0$ then $g_i \in \mathcal{U}_i$ has $\ell(g_i) = n_{i+1} \le n$ and so we are done by induction. If $j \ne 0$ then $\ell(g_i) = 1 + n_{i+1} \le n + 1$. If $\ell(g_i)$ happens to be $\le n$ then we are done by induction. If $\ell(g_i) = n + 1$ then we are done by the previous two cases. $\square$

The construction given above was first presented by Gupta and Sidki [GuSi83]; it is inspired by a construction of Grigorchuk [Gr80]. The first examples of finitely generated, infinite torsion groups were found in the 1960s; a discussion of the first known example can be found in [Ol95]. The example presented here does not completely settle all questions that can be derived from the problem Burnside posed in 1902. In fact, there is a stronger version, often called the *Burnside Problem*, which asks:

*If n is a fixed positive integer and G is a finitely generated group where every element in G satisfies $g^n = e$, is G finite?*

Burnside noted that this problem is quite tractable for $n = 2$, where the answer is "yes," and established that the claim is also true for $n = 3$.

Burnside's question can be further refined by asking if for given integers $k$ and $n$ (both $\geq 2$) there is an infinite group with a generating set of size $k$ where every element satisfies $g^n = e$. In fact, there is a "free Burnside group" $B(k, n)$ which is the quotient of the free group of rank $k$, $\mathbb{F}_k$, by the smallest normal subgroup containing the $n$th powers: $\{x^n \mid x \in \mathbb{F}_k\}$. There is an infinite group with generating set of size $k$ where every element satisfies $g^n = e$ if and only if $B(k, n)$ is infinite.

Burnside proved that $B(k, 3)$ is finite, for all $k$, but did not determine the order of this group. In 1933 Levi and van der Waerden proved that the order of $B(k, 3)$ is $3^n$ where

$$n = k + \binom{k}{2} + \binom{k}{3} .$$

(Question: Why does the fact that the group $B(2, 3)$ has order $3^3$ not contradict our claim that the group $\mathcal{U} < \text{Sym}(\mathcal{T})$, generated by $\sigma$ and $\omega$, is infinite?)

In 1940 Sanov showed that $B(k, 4)$ is always finite, and in 1958 Marshall Hall showed that the groups $B(k, 6)$ are all finite and was able to give a concrete formula for the size of $B(k, 6)$, similar to the formula above. It is still an open question if $B(2, 5)$ is finite or infinite.

On the other hand, for large values of $n$ there are finitely generated infinite groups where every element satisfies $g^n = e$. This was first shown by Novikov and Adian in 1968, who showed that $B(k, n)$ is infinite when $k > 2$ and $n$ is an odd number $\geq 665$.

There is yet another refinement, formulated by Wilhelm Magnus in the 1930s:

*For a fixed choice of k and n, is there is a largest, finite k generator group in which every element satisfies $g^n = e$?*

A rephrasing of this question asks if $B(k, n)$ has only finitely many finite quotients. In 1991 Efim Zelmanov showed that this so-called *restricted Burnside Problem* is true; this is work that earned him a Fields Medal in 1994.

**Exercises**

(1) Prove that the following groups are infinite torsion groups, but neither is finitely generated.

    a. A direct sum of a countable number of copies of $\mathbb{Z}_2$:

$$\mathbb{Z}_2 \oplus \mathbb{Z}_2 \oplus \mathbb{Z}_2 \oplus \cdots .$$

    The elements of this group consist of all $\mathbb{N}$-tuples of elements of $\mathbb{Z}_2$, where only finitely many entries are non-zero.

    b. The factor group $\mathbb{Q}/\mathbb{Z}$.

(2) Show that any finitely generated group $G$, where every element satisfies $g^2 = e$, is a finite group.

(3) Prove Case 2 of Lemma 6.7.

(4) Show that $\omega\sigma$ has order 9.

(5) Let $\mathcal{T}$ be the rooted, infinite binary tree and let $\text{SYM}(\mathcal{T})$ be the group of all symmetries of $\mathcal{T}$.

    a. Show that any $\alpha \in \text{SYM}(\mathcal{T})$ sends the root to the root, even if you just view $\mathcal{T}$ as an unrooted tree.

    b. Show that $\text{SYM}(\mathcal{T})$ contains an index 2 subgroup isomorphic to $\text{SYM}(\mathcal{T}) \oplus \text{SYM}(\mathcal{T})$.

    c. Show that $\text{SYM}(\mathcal{T})$ is uncountable. (Hence, by Exercise 11 in Chapter 1, $\text{SYM}(\mathcal{T})$ is not finitely generated.)

(6) Does $\text{SYM}(\mathcal{T})$ contain any element of infinite order?

(7) Let $\mathcal{T}_k$ be the regular, unrooted, $k$-valent tree, where $k \geq 3$. Show that $\text{SYM}(\mathcal{T}_k)$ is not finitely generated. (Hint: See Exercise 5.)

(8) Let $\mathcal{U}$ be the group constructed in this chapter.

    a. Prove that for any $g \in \mathcal{U}$, $g \neq e$, there is a finite group $H$ and a homomorphism $\phi : \mathcal{U} \to H$ such that $\phi(g) \neq e \in H$. (Hint: The group $\mathcal{U}$ acts on the set of vertices at level $k$.)

    b. Prove that the intersection of all finite index subgroups of $\mathcal{U}$ is the trivial subgroup $\{e\}$.

# 7

# Regular Languages and Normal Forms

The concept of a finite-state automaton has emerged as significant in many branches of human knowledge and understanding, including linguistics, computer science, philosophy, biology and mathematics .... In our work, finite-state automata are of fundamental importance: our objective is to use them to understand individual groups.

–from *Word Processing in Groups* [E+92]

## 7.1 Regular Languages and Automata

We have explored the word problem about as far as is prudent without being a bit more careful with words like "decide" and "construct." Here we introduce a standard mathematical model of computation. In thinking about modelling computation, there are a few key aspects that need to be preserved. One needs to have a way of inputing a sequence of information, and then a machine needs to execute various commands based on the input. One expects the total number of computations to be finite and possibly that there is some provision for the machine to have memory. The notion of finite-state automata, and their corresponding languages, fits these expectations.

**Definition 7.1.** Given an alphabet $\mathcal{S} = \{x_1, \ldots, x_n\}$, any subset of words in the free monoid on $\mathcal{S}$, $\mathcal{L} \subset \mathcal{S}^*$, is a *language*.

There are numerous examples of languages. If $\mathcal{S} = \{a\}$, then the set of all strings of even length, $\{a^{2n} \mid n \in \mathbb{N}\}$, forms a language. Similarly if $\mathcal{S}$ consists of the standard letters in the English alphabet, then the entries of any edition of the Oxford English Dictionary form a language.

The languages we are most interested in are those that can be produced by a somewhat simple computer.

**Definition 7.2.** An *automaton* consists of a directed graph $\mathcal{M}$, an associated alphabet $\mathcal{S}$, and a number of decorations:

1. There is a subset of vertices called the *start states*.
2. There is a subset $\mathcal{A} \subset V(\mathcal{M})$ called the *accept states*.
3. The directed edges are labelled by elements of the alphabet $\mathcal{S}$.

The *language* accepted by an automaton $\mathcal{M}$ is the set of all words $\omega \in \mathcal{S}^*$ corresponding to directed paths $p_\omega$ that begin at a start state and end at an accept state of $\mathcal{M}$. This language is denoted $L(\mathcal{M})$.

If $\mathcal{M}$ is finite then it is referred to as a *finite-state* automaton, which is commonly abbreviated as "FSA."

**Example 7.3.** The language $\{a^{3n} \mid n \in \mathbb{N}\}$ is the language associated to the automaton shown in Figure 7.1. As will be our convention in the rest of this book, vertices with double circles are the accept states, and vertices with an inscribed "S" are the start states.

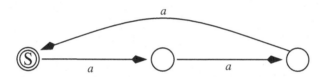

Fig. 7.1. A finite-state automaton whose language consists of the words $\{a^{3n} \mid n \geq 0\}$.

**Example 7.4.** Consider the finite-state automaton shown in Figure 7.2, whose associated alphabet is $\{a, b\}$. The vertices represented by double circles are the accept states, and the start state is indicated by the letter S. The language for this FSA consists of all words in $\{a, b\}^*$ that contain an even number of $b$'s.

**Example 7.5.** Every element in the infinite dihedral group $D_\infty$ can be represented uniquely by a word in the generators $\{a, b\}$ that alternates between $a$ and $b$. This set of words is the language accepted by either of the finite-state automata illustrated in Figure 7.3.

The FSA on the left in Figure 7.3 contains the FSA on the right. The additional vertex in the FSA on the left is called a *fail state*; once you reach it, you cannot leave it, and it is not an accept state. The FSA on the left also has a property that the FSA on the right does not. Namely, every word in $\{a, b\}^*$ can be traced along the edges of the FSA

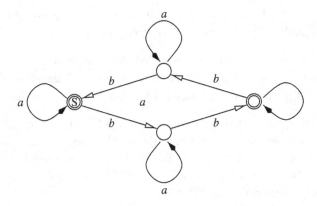

Fig. 7.2. A finite-state automaton whose language consists of all words in $\{a, b\}^*$ that contain an even number of $b$'s.

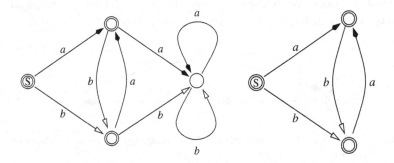

Fig. 7.3. Two automata whose languages are the reduced words describing elements in $D_\infty$.

on the left. In the FSA on the right, words containing $aa$ and $bb$ do not correspond to directed paths.

**Definition 7.6.** A *non-deterministic* automaton is an automaton where some edges are labelled by a new letter, $\varepsilon$, which is not in the original alphabet of the automaton. The *language* of a non-deterministic automaton is the set of all words corresponding to paths that begin at a start state and end at an accept state, with all of the $\varepsilon$'s stripped out.

**Example 7.7.** Let $\mathcal{L}$ be the language in $\{a, b, c\}^*$ consisting of words starting with a non-empty sequence of $a$'s, followed by a possibly empty string of $b$'s, followed by a possibly empty string of $c$'s. This language is the language of the non-deterministic automaton shown in Figure 7.4.

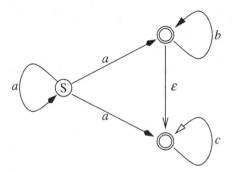

Fig. 7.4. A non-deterministic automaton whose language consists of all words in $\{a, b, c\}^*$ that begin with some $a$'s, followed by $b$'s and then $c$'s, where there may be no $b$'s and/or $c$'s.

**Definition 7.8.** A *deterministic* automaton, often abbreviated "DFA," is a finite-state automaton with the following additional requirements:

1. There is exactly one start state.
2. No two edges leaving a vertex have the same label.

A deterministic automaton is *complete* if for each vertex $v$, and each letter $a$ in the alphabet $S$, there is an edge leaving $v$ labelled $a$.

The assumption that there is exactly one start state is not always part of the definition of DFAs. As we will see in Theorem 7.12, in terms of the languages accepted by DFAs, the two conventions are equivalent. Similarly, the following result indicates that one can work with DFAs or complete DFAs and, in terms of the resulting languages, there is no difference. The proof of Lemma 7.9 is left as an end-of-chapter exercise.

**Lemma 7.9.** *If $\mathcal{L}$ is the language associated to a deterministic automaton, then $\mathcal{L}$ is the language associated to a complete, deterministic automaton.*

The following definition is sometimes stated as a theorem, the class of regular languages having been defined in terms of certain properties. However, for our purposes this definition is the most handy.

**Definition 7.10.** A *regular language* is any language that is accepted by a deterministic automaton.

**Theorem 7.11.** *Let $K$ and $L$ be regular languages with a common alphabet $S$. Then the following languages are also regular languages.*

1. *The complementary language* $\mathcal{S}^* - K$
2. *The union* $K \cup L$
3. *The intersection* $K \cap L$
4. *The concatenation* $KL$ *consisting of all words of the form*

$$KL = \{\omega_K \omega_L \mid \omega_K \in K \text{ and } \omega_L \in L\}$$

5. $K^* = K \cup KK \cup KKK \cup \cdots$

This theorem is most easily established using the following, somewhat surprising, fact.

**Theorem 7.12.** *The set of languages accepted by non-deterministic finite-state automata is the same as the set of regular languages.*

Theorem 7.12 is one of many results whose theme is "you may assume your automaton has properties X, Y, and Z." Such statements are often quite helpful in proving particular results about regular languages. However, the arguments establishing these results are not critical for understanding the applications to group theory.

Before starting the proof of Theorem 7.12, we consider how one would test if a given word is in the language of a non-deterministic automaton. It is clear how to test if a word is in the language of a deterministic automaton. Every word corresponds to a unique path in the DFA, and so you just follow the associated path and check if it ends in an accept state. But in a non-deterministic automaton there are choices, and we have to keep track of these choices. If you have a large stack of paper available, then you could test a given word $\omega$ by making piles, one pile per start state in the given automaton $\mathcal{M}$. Then, reading off the first letter of $\omega$, you would consider each start state, and write down all of the locations that you can arrive at from that start state, using only the first letter of $\omega$ and any available $\varepsilon$-edges. You then consult your lists, and make a new set of stacks, one stack of paper for each state you could be in after reading the first letter of $\omega$. Continue to follow this process until you have read all of $\omega$. If the resulting set of states contains an accept state, then there is a path from some start state to some accept state, which is labelled by the word $\omega$ (with possibly some $\varepsilon$'s tossed in along the way). The proof of Theorem 7.12 follows this line of thought.

*Proof of Theorem 7.12.* It is clear that every deterministic FSA is a non-deterministic FSA, so the set of regular languages is contained in the set of languages accepted by non-deterministic FSAs. The trick is proving the other containment.

Let $\mathcal{M}$ be a non-deterministic finite-state automaton. We first build an automaton that accepts the same language as $\mathcal{M}$ and that does not use the edge label $\varepsilon$. Let $S$ be a subset of $V(\mathcal{M})$. The $\varepsilon$-*closure* of $S$, $\varepsilon(S)$, is the set of all vertices of $\mathcal{M}$ that can be reached by a directed path beginning in $S$. A subset $S \subset V(\mathcal{M})$ is $\varepsilon$-*closed* if $S = \varepsilon(S)$.

Let $\mathcal{M}_\varepsilon$ be an automaton whose states are the $\varepsilon$-closed subsets of $V(\mathcal{M})$. If $S \subset V(\mathcal{M})$, and $a \in S$, define $Sa$ to be the set of all states at the ends of edges labelled $a$ that begin in $S$. The start states of $\mathcal{M}_\varepsilon$ are all $\varepsilon$-closed subsets that contain a start state; the accept states are all $\varepsilon$-closed subsets that contain an accept state of $\mathcal{M}$; and there is an edge labelled $a$ from $S$ to $\varepsilon(Sa)$ for each $a \in A$. By construction, in $\mathcal{M}_\varepsilon$ there are no edges labelled by $\varepsilon$, and the reader can verify that $L(\mathcal{M}) = L(\mathcal{M}_\varepsilon)$.

We can now convert $\mathcal{M}_\varepsilon$ into a deterministic automaton, $\mathcal{D}$. The states of $\mathcal{D}$ consist of all the subsets of $V(\mathcal{M}_\varepsilon)$. The single start state of $\mathcal{D}$ is the subset of $V(\mathcal{M}_\varepsilon)$ consisting of all the start states of $\mathcal{M}_\varepsilon$. The accept states of $\mathcal{D}$ are the subsets of $V(\mathcal{M}_\varepsilon)$ that contain at least one accept state of $\mathcal{M}_\varepsilon$. In $\mathcal{D}$ there is an edge from $U$ to $U'$ labelled by $x$ if, for each $v \in U$, there is an edge labelled $x$ from $v$ to some $v' \in U'$, and $U'$ is entirely composed of such vertices. That is, there is an edge labelled $x$ from $U$ to the vertex corresponding to the set

$$U' = \{v' \in V(\mathcal{M}_\varepsilon) \mid v' \text{ is at the end of an edge}$$
$$\text{labelled } x \text{ that begins at some } v \in U\}.$$

By its construction, $\mathcal{D}$ is a deterministic automaton. Further, if $\omega \in L(\mathcal{M}_\varepsilon)$ then $\omega$ describes a directed edge path in $\mathcal{D}$ starting at the start state. Further, the terminal vertex of the associated path $p_\omega$ corresponds to a set of vertices of $\mathcal{M}_\varepsilon$, at least one of which is an accept state of $\mathcal{M}_\varepsilon$, since $\omega \in L(\mathcal{M}_\varepsilon)$. Thus $L(\mathcal{D}) = L(\mathcal{M}_\varepsilon) = L(\mathcal{M})$.     $\square$

*Proof of Theorem 7.11.* (1) By Lemma 7.9 we may assume that the language $K = L(\mathcal{M})$ where $\mathcal{M}$ is a complete deterministic automaton. The language $\mathcal{S}^* - K$ is then accepted by the machine $\mathcal{M}^\circ$ formed by taking $\mathcal{M}$ and making all non-accept states into accept states, and making all accept states non-accept states.

(2) Let $\mathcal{M}_K$ and $\mathcal{M}_L$ be deterministic automata whose languages are $K$ and $L$ respectively. The disjoint union of $\mathcal{M}_K$ and $\mathcal{M}_L$ is a non-deterministic automaton (with two start states) that accepts $K \cup L$. By Theorem 7.12, $K \cup L$ is then regular.

(3) Since $K \cap L = \mathcal{S}^* - ((\mathcal{S}^* - K) \cup (\mathcal{S}^* - L))$, this follows from (1) and (2).

(4) Construct a non-deterministic automaton by forming the disjoint union of $\mathcal{M}_K$ and $\mathcal{M}_L$, declaring that the only start state of the new machine is the start state of $\mathcal{M}_K$, and adding edges labelled $\varepsilon$ from each accept state of $\mathcal{M}_K$ to the start state of $\mathcal{M}_L$.

(5) From each accept state in $\mathcal{M}_K$, add a $\varepsilon$-labelled edge connecting it to the start state of $\mathcal{M}_K$. $\qquad\square$

**Example 7.13.** The set of freely reduced words in $\{x, x^{-1}, y, y^{-1}\}^*$ forms a regular language. An automaton accepting this language is shown in Figure 7.5. It has five states, one for each possible "last letter read" as one considers a given word. Initially there is no "last letter read" and so one starts at the central vertex, which is connected to each of the other four states. If one is in the state corresponding to having just read an $x$, then one can transition to any state *except* the state corresponding to having just read an $x^{-1}$, as one would not read $x^{-1}$ after an $x$ in a freely reduced word. The rest of the structure of this FSA is similar.

The elements of the free group $\mathbb{F}_2$ correspond to the freely reduced words in $\{x, x^{-1}, y, y^{-1}\}^*$. Thus these elements correspond to the words in a regular language. The even subgroup of $\mathbb{F}_2$, introduced in the proof of Proposition 3.12, corresponds to all freely reduced words that are of even length. It is easy to construct an FSA whose language consists of all words in $\{x, x^{-1}, y, y^{-1}\}^*$ that are of even length. (Try it!) Thus by part (3) of Theorem 7.11, the set of words corresponding to the even subgroup is also a regular language.

## 7.2 Not All Languages are Regular

Not all languages are regular languages. In Example 7.13 a deterministic automaton is constructed that can remember a finite amount of information. Namely, the machine can remember the last letter that it has read. One of the easiest ways to detect if a language is regular or not is to consider if one would need to have unlimited memory in order to determine if an arbitrary word is in the language. This instinct is codified by the Pumping Lemma.

**Theorem 7.14** (Pumping Lemma). *Let $\mathcal{L}$ be a regular language. Then there is an integer $n \geq 1$ such that any word $x \in \mathcal{L}$ of length greater than $n$ can be expressed as $x = uvw$ where $v$ is a non-empty word, and*

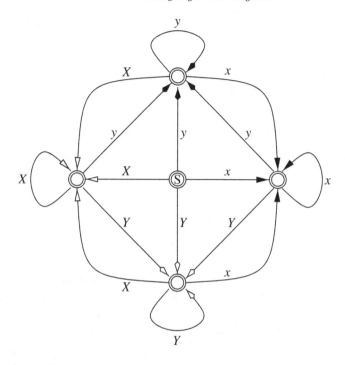

Fig. 7.5. A finite-state automata whose language consists of all freely reduced words in $\{x, x^{-1}, y, y^{-1}\}^*$. Here we have used $X$ to denote $x^{-1}$ and $Y$ to denote $y^{-1}$.

1. $|u| < n$;

2. $uv^i w \in \mathcal{L}$ for all $i \geq 0$.

*Proof.* Let $\mathcal{M}$ be a complete DFA whose associated language is $\mathcal{L}$. Set $n = |V(\mathcal{M})|$ and let $x$ be a word of length greater than $n$. Because $\mathcal{M}$ is a complete DFA, there is a unique directed edge path starting at the start state whose sequence of labels matches the word $x$. Since the length of $x$ is greater than $n$, this path must visit some vertex $z \in V(\mathcal{M})$ twice. Thus the path associated to $x$ contains a circuit. Let $u$ be the (possibly empty) prefix of $x$ associated to the initial portion of the path from the start state to $z$; let $v$ be the subword corresponding to the subpath from $z$ back to $z$; and let $w$ be the (possibly empty) suffix of $x$ corresponding to the part of the path from $z$ to the accept state at the end of this path. It then follows that the words $uv^i w$ (for $i \geq 0$) all describe paths starting at the start state and ending at an accept state. The only difference in the paths is how many times they wind around the closed path from $z$ to $z$. □

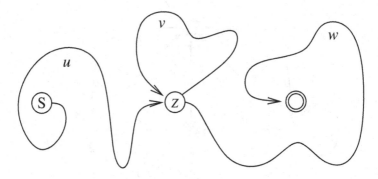

Fig. 7.6. The idea behind the Pumping Lemma.

**Example 7.15.** The Pumping Lemma can be used to prove that the language $\mathcal{L} = \{a^n b^n \mid n \geq 1\}$ is not a regular language. Assume to the contrary that $\mathcal{L}$ is regular and let $n$ be the number of states of an automaton $\mathcal{M}$ whose language is $\mathcal{L}$. Then the path corresponding to $a^{n+1}$ contains a closed subpath, of some length $k$. It follows that the word $a^{n+1+k} b^n$ must also be accepted by $\mathcal{M}$, contradicting the claim that $\mathcal{L} = L(\mathcal{M})$.

**Definition 7.16.** Let $\mathcal{L}$ be a language in $\mathcal{S}^*$ and let $\omega \in \mathcal{S}^*$. The *cone type* of $\omega$ is the set of all words $\omega' \in \mathcal{S}^*$ such that $\omega \cdot \omega' \in \mathcal{L}$. Denote the cone type of $\omega$ by $\mathrm{Cone}(\omega)$. The collection

$$\mathrm{Cone}(\mathcal{L}) = \{\mathrm{Cone}(\omega) \mid \omega \in \mathcal{S}^*\}$$

is the *cone type of the language* $\mathcal{L}$.

**Example 7.17.** Let $\mathcal{S} = \{a\}$ and let $\mathcal{L}$ be the language of all words of even length. Then $\mathrm{Cone}(a) = \mathrm{Cone}(a^3) = \mathrm{Cone}(a^5) = \cdots$ and they are all equal to the set of words of odd length, while $\mathrm{Cone}(\epsilon) = \mathrm{Cone}(a^2) = \mathrm{Cone}(a^4) = \cdots$ are all the words of even length. Thus $\mathrm{Cone}(\mathcal{L})$ consists of exactly two sets, the set of words of even length and the set of words of odd length.

**Theorem 7.18** (Myhill–Nerode Theorem). *A language $\mathcal{L}$ is regular if and only if its set of cone types is finite.*

*Proof.* We first consider the forward direction. If $\mathcal{L}$ is regular then it is accepted by a complete DFA $\mathcal{M}$. For each vertex $v \in V(\mathcal{M})$ let $\mathcal{M}_v$ be the automaton formed by making $v$ the only start state. The cone type of a given word $\omega$ is then determined by the end vertex of the associated path $p_\omega$. If this vertex is denoted $v_\omega$ then $\mathrm{Cone}(\omega) = L(\mathcal{M}_{v_\omega})$. Thus the number of cone types is at most the number of vertices in $\mathcal{M}$.

To establish the converse, assume that $\mathrm{Cone}(\mathcal{L})$ is finite. We construct an automaton $\mathcal{M}$ starting with declaring the states of $\mathcal{M}$ to be the cone types of $\mathcal{L}$. Designate $\mathrm{Cone}(\varepsilon)$ to be the unique start state, where $\varepsilon$ represents the empty word. The accept states of $\mathcal{M}$ are the cone types that contain $\varepsilon$.

We add an edge labelled $x$ from the vertex given by $\mathrm{Cone}(\omega)$ to $\mathrm{Cone}(\omega x)$ for each cone type and each $x \in \mathcal{S}$. There is a possibility that this definition depends on the word $\omega$, not on the cone type of $\omega$. That is, perhaps $\mathrm{Cone}(\omega) = \mathrm{Cone}(\widehat{\omega})$, but $\mathrm{Cone}(\omega \cdot x) \neq \mathrm{Cone}(\widehat{\omega} \cdot x)$. This cannot happen, because if $x \in \mathcal{S}$ and $\omega \in \mathcal{S}^*$ then

$$\mathrm{Cone}(\omega x) = \{\omega' \in \mathcal{S}^* \mid x\omega' \in \mathrm{Cone}(\omega)\} .$$

Thus if $\mathrm{Cone}(\omega) = \mathrm{Cone}(\widehat{\omega})$ then

$$\mathrm{Cone}(\omega x) = \{\omega' \in \mathcal{S}^* \mid x\omega' \in \mathrm{Cone}(\omega)\}$$
$$= \{\omega' \in \mathcal{S}^* \mid x\omega' \in \mathrm{Cone}(\widehat{\omega})\} = \mathrm{Cone}(\widehat{\omega}x)$$

for each $x \in \mathcal{S}$. The reader may check that this automaton $\mathcal{M}$ is a complete DFA.

What is the language accepted by $\mathcal{M}$? First, we can characterize $\mathcal{L}$ by noticing that a word $\omega$ is in $\mathcal{L}$ if and only if the empty word $\varepsilon \in \mathrm{Cone}(\omega)$. Let $\omega$ be an arbitrary word in $\mathcal{S}^*$. Since $\mathcal{M}$ is a complete DFA, there is a unique path in $\mathcal{M}$, starting at the start state, and traversing edges as described by $\omega$. By the construction of $\mathcal{M}$, this path ends at the state associated to $\mathrm{Cone}(\omega)$. Hence $\omega \in L(\mathcal{M})$ if and only if $\mathrm{Cone}(\omega)$ is an accept state, which is if and only if $\varepsilon \in \mathrm{Cone}(\omega)$. Thus $L(\mathcal{M}) = \mathcal{L}$ by the characterization of $\mathcal{L}$ given at the start of this paragraph. □

**Example 7.19.** Let the alphabet be $\mathcal{S} = \{a, b\}$ and let

$$\mathcal{L} = \{\omega\omega \mid \omega \in \mathcal{S}^*\}$$

be the language of formal squares of words in $\mathcal{S}^*$. Then $ba^i b \in \mathrm{Cone}(a^i)$ but $ba^i b \notin \mathrm{Cone}(a^j)$ for any $j \neq i$. Thus the language $\mathcal{L}$ has infinitely many distinct cone types, so, by the Myhill–Nerode Theorem, $\mathcal{L}$ is not regular.

### 7.3 Regular Word Problem?

We began our discussion of regular languages motivated by a desire to have a more formal understanding of what it means for something to be computable. We now return to the group theory setting, ready to test drive our new understanding.

Let $G$ be a finitely generated group with generating set $S$. Define

$$\mathrm{WP}(G,S) = \{\omega \in \{S \cup S^{-1}\}^* \mid \pi(\omega) = e \in G\}.$$

Thus $\mathrm{WP}(G,S)$ is the set of all words in the generators and their inverses that evaluates to the identity in $G$. This set is often colloquially referred to as "the word problem" for $G$, since being able to solve Dehn's word problem for $G$ (and generating set $S$) amounts to being able to determine if a given word $\omega \in \{S \cup S^{-1}\}^*$ is in $\mathrm{WP}(G,S)$.

Since $\mathrm{WP}(G,S) \subset \{S \cup S^{-1}\}^*$, it is a language, and one can ask: *When is $\mathrm{WP}(G,S)$ a regular language?* Surprisingly, it is not easy to come up with examples where the word problem is a regular language. For example, let $G$ be an infinite cyclic group with generator $x$. Then a word in $\{x, x^{-1}\}^*$ represents the identity if and only if the total exponent sum is zero. But then $x^{-i} \in \mathrm{Cone}(x^i)$ but $x^{-i} \notin \mathrm{Cone}(x^j)$ for any $i$ and $j \in \mathbb{Z}$ where $j \neq i$. So $\mathrm{WP}(G,x)$ has infinitely many cone types, and the Myhill–Nerode Theorem (Theorem 7.18) implies that $\mathrm{WP}(G,x)$ is not a regular language. Faced with this non-example, you might start looking at free groups, free products, the Baumslag–Solitar group $\mathrm{BS}(1,2)$, etc., but you would only be wasting your time.

**Theorem 7.20.** *The word problem $\mathrm{WP}(G,S)$ is a regular language if and only if $G$ is a finite group.*

*Proof.* If $G$ is finite, the Cayley graph with respect to $\{S \cup S^{-1}\}$ forms a complete DFA, with start and accept state the vertex corresponding to the identity. The language accepted by this DFA is precisely $\mathrm{WP}(G,S)$.

Conversely, suppose $\mathrm{WP}(G,S) = L(\mathcal{M})$ for some finite-state automaton $\mathcal{M}$. One may assume that every vertex in $\mathcal{M}$ is connected by a directed path to an accept state. Otherwise form a new automaton that accepts the same language as $\mathcal{M}$ by simply removing that vertex and all incident edges. Repeatedly applying this technique trims $\mathcal{M}$ to a finite automaton where every vertex is connected by a directed path to at least one accept state.

Let $\omega$ and $\omega'$ be two words whose corresponding paths both end at $z \in \mathcal{M}$. Let $v$ be a word describing a path from $z$ to an accept state of

$\mathcal{M}$. Then $\omega v$ and $\omega' v$ are both in $L(\mathcal{M})$. It follows that

$$\pi(\omega v) = \pi(\omega' v) = e \in G.$$

But in the group $G$ we have $\pi(\omega)\pi(v) = \pi(\omega')\pi(v)$. Thus, by right cancellation, $\pi(\omega) = \pi(\omega')$. So if two words describe paths ending at the same vertex, then they represent the same element of $G$. But then the order of $G$ can be no larger than the number of states of the finite-state automaton of $\mathcal{M}$. $\qquad\square$

**Remark 7.21.** Theorem 7.20 was first proved by Anisimov in 1971 [An71]. It is a striking result that seems to show that asking when $\mathrm{WP}(G, S)$ is a regular language is not the right question, especially if one is mainly interested in infinite groups!

Readers with a background in formal language theory might be interested in exploring an analogous theorem of Muller and Schupp, who showed that $\mathrm{WP}(G, S)$ is a context-free language if and only if $G$ contains a free subgroup of finite index [MuS83].

## 7.4 A Return to Normal Forms

Recall that, given a group $G$ and generating set $S$, a normal form is a subset of the free monoid $\{S \cup S^{-1}\}^*$ that maps bijectively to $G$ under the evaluation map $\pi : \{S \cup S^{-1}\}^* \twoheadrightarrow G$. (This idea was introduced in Section 5.1.) Thus a normal form $\mathcal{NF} \subset \{S \cup S^{-1}\}^*$ can be thought of as a language, and one might hope that it forms a regular language. When this happens, we say $G$ has a *regular normal form*. It is also common to say that $G$ *admits* a regular normal form.

**Example 7.22.** The freely reduced words are a normal form for $\mathbb{F}_n$. As part of Example 7.13 we showed that this is a regular language, hence $\mathbb{F}_n$ has a regular normal form.

**Example 7.23.** Let $a$ and $b$ be generators of the free abelian group $\mathbb{Z} \oplus \mathbb{Z}$. Then every element can be expressed as $a^i b^j$ for some $i$ and $j \in \mathbb{Z}$. (This is $\mathrm{NF}_1$ from Example 5.2.) This too is a regular language. Perhaps this is most easily shown by noting that $\{a^i \mid i \in \mathbb{Z}\}$ and $\{b^j \mid j \in \mathbb{Z}\}$ are regular languages, and then appeal to the fact that the concatenation of two regular languages forms a regular language.

**Proposition 7.24.** *Let $S$ be a finite generating set for a group $G$. If $G$ has a regular normal form $\mathcal{N}(G) \subset \{S \cup S^{-1}\}^*$ then for any other finite generating set $T$ there is a regular normal form contained in $\{T \cup T^{-1}\}^*$.*

*Proof.* Since $T$ is a set of generators, each $s \in S$ (or $S^{-1}$) can be represented as a word in $\{T \cup T^{-1}\}^*$, $s = \omega_s$. The language obtained by replacing each generator $s$ by $\omega_s$ in each word in $\mathcal{N}(G)$ is then a regular normal form in $\{T \cup T^{-1}\}^*$. The corresponding automaton is formed by replacing each edge labelled $s$ by an edge path labelled $\omega_s$, where none of the new vertices are start or accept states. $\quad\square$

**Example 7.25.** The infinite dihedral group $D_\infty$ is generated by two reflections, $a$ and $b$. It is also generated by a reflection, $a$, and a translation $\tau = ba$. A regular normal form for $D_\infty$ with respect to $\{a, b\}$ consists of all words which alternate $a$'s and $b$'s, and an FSA accepting this language is shown on the left in Figure 7.7. To construct a normal

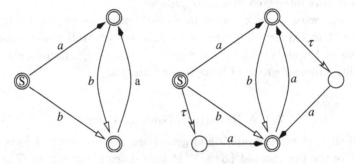

Fig. 7.7. Converting the finite-state automaton giving a regular normal form for $D_\infty$ with respect to $\{a, b\}$ to one using $\{a, \tau, \tau^{-1}\}$ as its generating set.

form using $\{a, \tau, \tau^{-1}\}$ we note that $a = a$ and $b = \tau a$. Thus we should replace all edges labelled $b$ in the original FSA by edge paths labelled $\tau a$. We note that the resulting normal form is neither efficient nor natural to the new set of generators. For example, instead of using the word $\tau^{-1}$ to represent a horizontal translation to the left, this new normal form uses $a\tau a$.

The following is an exercise in applying Theorem 7.11.

**Proposition 7.26.** *If $G$ and $H$ have regular normal forms, then so does $G \oplus H$ and $G * H$.*

**Corollary 7.27.** *Any free product of two finite groups has a regular normal form.*

*Proof.* Every finite group admits a regular normal form. Simply take $S = G \setminus \{e\}$, and let $\mathcal{N}(G)$ be the words of length $\leq 1$. The result then follows from Proposition 7.26. $\quad\square$

Bob Gilman introduced the study of groups admitting regular normal forms in 1987 [Gi87], where he established the following result:

**Theorem 7.28.** *If $G$ has a regular normal form, and $G$ is not finite, then $G$ has an element of infinite order.*

*Proof.* Let $\mathcal{M}$ be a complete automaton whose language is a regular normal form for $G$, and let $n$ be the number of vertices of $\mathcal{M}$. Then since $G$ is infinite there is a word $x \in L(\mathcal{M})$ of length greater than $n$. So by the Pumping Lemma (7.14) we may write $x = uvw$ where, for each $i \geq 0$, $uv^iw \in L(\mathcal{M})$. Since $L(\mathcal{M})$ is a normal form, $\pi(uv^iw) \neq \pi(uv^jw)$ for $i \neq j$. But then

$$\pi(uv^iw) \neq \pi(uv^jw)$$
$$\Rightarrow \quad \pi(u)\pi(v^i)\pi(w) \neq \pi(u)\pi(v^j)\pi(w)$$
$$\Rightarrow \quad \pi(v)^i \neq \pi(v)^j$$

for $i \neq j$. It follows that $\pi(v)$ is an element of infinite order in $G$. $\qquad \square$

A mild restatement of this result is:

**Corollary 7.29.** *Finitely generated, infinite torsion groups, like the group $\mathcal{U}$ considered in Chapter 6, do not have regular normal forms.*

## 7.5 Finitely Generated Subgroups of Free Groups

If $G$ is a finitely generated infinite group, it may be the case that some subgroups of $G$ are not finitely generated. For example, as a homework problem in Chapter 3 you showed that the kernel of the map from $\mathbb{F}_2$ onto $\mathbb{Z} \oplus \mathbb{Z}$ is a free group with basis

$$S = \{[x^i, y^j] \mid i,j \in \mathbb{Z} \text{ and } i \cdot j \neq 0\}.$$

Here is another normal subgroup of $\mathbb{F}_2$ that is not finitely generated.

**Lemma 7.30.** *Let $\mathbb{F}_2$ be generated by $x$ and $y$ and let $\phi : \mathbb{F}_2 \to \mathbb{Z}$ be the map induced by sending both $x$ and $y$ to 1. The kernel of $\phi$ is free with basis $S = \{x^iy^{-i} \mid i \in \mathbb{Z}\}$.*

*Proof.* The given set is in the kernal, and the quotient by the normal subgroup containing these elements is isomorphic to $\mathbb{Z}$. So it suffices to show that the subgroup of $\mathbb{F}_2$ generated by $S$ is normal. But the following computations show that the conjugate of each element in $S$ is in the subgroup generated by $S$

$$
\begin{aligned}
x \cdot x^i y^{-i} \cdot x^{-1} &= x^{i+1} y^{-(i+1)} \cdot y x^{-1} = x^{i+1} y^{-(i+1)} \cdot (xy^{-1})^{-1}, \\
x^{-1} \cdot x^i y^{-i} \cdot x &= x^{i-1} y^{-(i-1)} \cdot y^{-1} x = x^{i-1} y^{-(i-1)} \cdot (x^{-1}y)^{-1}, \\
y \cdot x^i y^{-i} y^{-1} &= y x^{-1} \cdot x^{i+1} y^{-(i+1)} = (xy^{-1})^{-1} \cdot x^{i+1} y^{-(i+1)}, \\
y^{-1} \cdot x^i y^{-i} \cdot y &= y^{-1} x \cdot x^{i-1} y^{-(i-1)} = (x^{-1}y)^{-1} \cdot x^{i-1} y^{-(i-1)}.
\end{aligned}
$$

So the subgroup generated by $S$ is normal. Exercise 7 asks you to show that $S$ forms a basis. $\qquad\square$

The following example, combined with Theorem 7.32 below, indicates that determining which subgroups of a given group $G$ are finitely generated can be rather subtle.

**Example 7.31.** The intersection of two finitely generated subgroups of a group $G$ is not necessarily a finitely generated subgroup. Consider the group $\mathbb{F}_2 \oplus \mathbb{Z}$ with $\{x, y\}$ being the basis for $\mathbb{F}_2$ and $z$ a cyclic generator of $\mathbb{Z}$. Then the kernel $K$ of the map $\varphi : \mathbb{F}_2 \oplus \mathbb{Z} \to \mathbb{Z}$ induced by

$$
\varphi(x) = \varphi(y) = \varphi(z) = 1 \in \mathbb{Z}
$$

is a free group generated by $\{x^{-1}z, y^{-1}z\}$. The proof of this claim is similar to the proof of Lemma 7.30: show that the subgroup generated by this set is normal and note that the quotient is indeed $\mathbb{Z}$. Normality is established by equations of the form:

$$
y \cdot x^{-1} z \cdot y^{-1} = y z^{-1} \cdot x^{-1} z \cdot y^{-1} z.
$$

Working out the details is left to Exercise 13.

The intersection of $K$ with direct factor $\mathbb{F}_2$ is the kernel of the map $\phi : \mathbb{F}_2 \to \mathbb{Z}$ considered in Lemma 7.30 above:

$$
\ker(\varphi) \cap \mathbb{F}_2 = \ker(\phi).
$$

The subgroups $K = \ker(\varphi)$ and $\mathbb{F}_2$ are both finitely generated but, as we have seen, $\ker(\phi)$ is not finitely generated.

**Theorem 7.32** (Howson's Theorem). *The intersection of finitely generated subgroups of a free group is a finitely generated free group.*

Our proof of Howson's Theorem will exploit our understanding of regular languages. If $G$ is a group with finite generating set $S$, then a subgroup $H < G$ is the *image of a regular language* over $S \cup S^{-1}$ if there is a regular language $\mathcal{L} \subset \{S \cup S^{-1}\}^*$ such that $\pi(\mathcal{L}) = H$. This condition is independent of one's choice of finite generating set (see the analogous argument in the proof of Proposition 7.24). Moreover, there is a surprising characterization of which subgroups are the images of regular languages:

**Theorem 7.33.** *Let $S$ be a finite set of generators for a group $G$. A subgroup $H$ of $G$ is the image of a regular language $\mathcal{L} \subset \{S \cup S^{-1}\}^*$ if and only if $H$ is finitely generated.*

*Proof.* If $H$ is finitely generated, then fix a generating set $T$ and for each $t \in T$ choose a word $\omega_t \in \{S \cup S^{-1}\}^*$ such that $\pi(\omega_t) = t$. The free monoid on these words and their formal inverses is a regular language that projects to $H$ under $\pi$.

Conversely let $\mathcal{L} = L(\mathcal{M})$ be a regular language that projects to a subgroup $H < G$. We may assume that $\mathcal{M}$ has a number of nice properties:

1. There are no $\varepsilon$-edges.
2. There is a single start state that is not the terminal vertex of any directed edge of $\mathcal{M}$. This may or may not be an accept state. (Exercise)
3. Aside from the possibility of the start state being an accept state, there is a single accept state. This accept state is not the initial vertex of any directed edge of $\mathcal{M}$. (Exercise)
4. Every vertex $v$ of $\mathcal{M}$ can be reached by a directed path beginning at the start state (otherwise remove such vertices).
5. From every vertex $v$ of $\mathcal{M}$ there is a directed path from $v$ to the accept state of $\mathcal{M}$ (otherwise remove such vertices).

Given all these conditions, there is a directed spanning tree $T \subset \mathcal{M}$, which is rooted at the start state of $\mathcal{M}$. In other words, $T$ is a subdirected graph that contains every vertex of $M$ and, given any vertex, there is a path in $T$ from the single start state to that vertex. For each edge $e \in \mathcal{M} \setminus T$ let $\alpha(e)$ denote the label of the path in $T$ connecting the start state to the initial vertex of $e$, and let $\beta(e)$ be the label of the path connecting the start state to the terminal vertex of $e$. Let $z$ denote the label of the path in $T$ from the start state to the accept state.

If $\omega \in \mathcal{L}$ then $\omega$ describes a path from the start state of $\mathcal{M}$ to the accept state of $\mathcal{M}$. We may then view $\omega$ as a sequence of subwords

$$\omega = g_0 h_1 g_1 h_2 g_2 \cdots g_{n-1} h_n g_n$$

where each $g_i$ is the label of a path contained in the spanning tree $T$ and each $h_i$ is the label of an edge $e_i \notin T$. Note: while some of the $g_i$ may be empty words, we may assume that none of the $h_i$ are

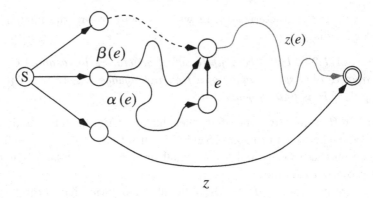

Fig. 7.8. The spanning tree and other featured elements of the automaton.

empty. It follows from the definitions that $\alpha(e_1) = g_0$ and more generally $\alpha(e_{i+1}) = \beta(e_i)g_i$. Thus we may write

$$\pi[\omega] = \pi[\alpha(e_1)h_1\beta(e_1)^{-1} \cdot \alpha(e_2)h_2\beta(e_2)^{-1} \cdots \alpha(e_n)h_n\beta(e_n)^{-1} \cdot z].$$

If we let $\ell(e)$ denote the label of an edge $e \in \mathcal{M}$ then the image of $\mathcal{L}$ is contained in the subgroup generated by

$$E = \{\pi(z), \pi[\alpha(e)\ell(e)\beta(e)^{-1}] \mid e \text{ is an edge in } \mathcal{M} \setminus \mathcal{T}\}.$$

It then remains to be shown that each of these elements is contained in $H$, for then $E$ will be a finite set of generators for $H$. Since $z$ ends in an accept state, $\pi(z) \in H$. If $e$ is an edge in $\mathcal{M} \setminus \mathcal{T}$ let $z(e)$ be a path from the target vertex of $e$ to the accept state. Then

$$\pi[\alpha(e)e\beta(e)^{-1}] = \pi[\alpha(e)ez(e) \cdot (\beta(e)z(e))^{-1}].$$

But $\pi[\alpha(e) \cdot e \cdot z(e)]$ and $\pi[(\beta(e) \cdot z(e))]$ are both in $H$, so $\pi[\alpha(e) \cdot e \cdot \beta(e)^{-1}]$ is in $H$. $\qquad\square$

The proof given above, when restricted to the case of subgroups of free groups, establishes the following result:

**Corollary 7.34.** *Finitely generated subgroups of free groups are free.*

(This result is, of course, an immediate corollary of the Nielsen–Schreier Theorem 3.23, but the formal language argument given below is useful none-the-less.)

*Proof.* If $H$ is a finitely generated subgroup of $\mathbb{F}_n$ then one direction of Theorem 7.33 implies that $H$ is the image of a regular language $\mathcal{L}$. The proof of Theorem 7.33 indicates that every $h \in H$ can be expressed as a product of the elements

$$E = \{\pi(z), \pi[\alpha(e)\ell(e)\beta(e)^{-1}] \mid e \text{ is an edge in } \mathcal{M} \setminus \mathcal{T}\}$$

and their inverses. But any such word which is freely reduced when written as a product of elements of $E \cup E^{-1}$ cannot represent the identity in $\mathbb{F}_n$. Consider for example the product

$$\alpha(e_1)\ell(e_1)\beta(e_1)^{-1} \cdot \left(\alpha(e_2)\ell(e_2)\beta(e_2)^{-1}\right)^{-1}$$
$$= \alpha(e_1)\ell(e_1)\beta(e_1)^{-1} \cdot \beta(e_2)\ell(e_2)^{-1}\alpha(e_2).$$

If $\beta(e_1)^{-1} \cdot \beta(e_2)$ cancels completely, that means that the initial vertex of $e_2$ is the initial vertex of $e_1$. But since this expression was freely reduced when written in $E \cup E^{-1}$, $e_1 \neq e_2$. Thus no more cancellation occurs. An induction argument along these lines establishes that a freely reduced word written with $n$ letters from $E \cup E^-$ is equivalent to a freely reduced word of length at least $n$, when written in terms of the generators of $\mathbb{F}_n$. Thus

$$E = \{\pi(z), \pi[\alpha(e)\ell(e)\beta(e)^{-1}] \mid e \text{ is an edge in } \mathcal{M} \setminus \mathcal{T}\}$$

forms a free basis for the image of $\mathcal{L}$, and so $H$ is a free subgroup. □

Before finally proving Howson's Theorem, we have one more lemma that needs to be established.

**Lemma 7.35.** *Let $\mathcal{L}$ be a regular language over $\{S \cup S^{-1}\}$ and let $\mathcal{R}$ be the language formed by freely reducing the words in $\mathcal{L}$. If $\mathcal{L}$ is regular, then so is $\mathcal{R}$.*

*Proof.* Let $\mathcal{L} = L(\mathcal{M})$, for some finite automaton $\mathcal{M}$. If for any $s \in S$ there is a path in $\mathcal{M}$ with label $ss^{-1}$ or $s^{-1}s$ (some edges of which may be labelled by $\varepsilon$) from a vertex $u$ to a vertex $v$, add a new edge from $u$ to $v$ labelled $\varepsilon$. Call this new automaton $\mathcal{M}'$. If $\mathcal{M}$ accepted a not-freely reduced word $wss^{-1}u$ then the new automaton will accept this word and the word $wu$. Thus $\mathcal{M}'$ accepts both $\mathcal{L}$ and the reduced words formed from words in $\mathcal{L}$. Intersecting $L(\mathcal{M}')$ with the regular language of freely reduced words, and applying Theorem 7.11, establishes the lemma. □

*Proof of Howson's Theorem.* Let $H$ and $K$ be two finitely generated subgroups of $\mathbb{F}_n$. Then $H$ is the image of a regular language $\mathcal{L}_H$ and $K$ is the image of a regular language $\mathcal{L}_K$. By Lemma 7.35 we may assume both languages consist entirely of freely reduced words, which since $\mathbb{F}_n$

is free means they consist of all freely reduced words representing elements of $H$ and $K$. The intersection $\mathcal{L}_H \cap \mathcal{L}_K$ is then a regular language consisting of all freely reduced words which represent elements in $H$ and in $K$, hence $\mathcal{L}_H \cap \mathcal{L}_K$ maps onto $H \cap K$. So $H \cap K$ is finitely generated by Theorem 7.33.                                    □

In the argument above, being able to claim that the languages consist of freely reduced words is critical. In general, if $\mathcal{L}_H$ and $\mathcal{L}_K$ are two regular languages whose images are $H$ and $K$ respectively, then $\mathcal{L}_H \cap \mathcal{L}_K$ maps to $H \cap K$, but it does not have to map onto $H \cap K$. There may be elements in the intersection that have different words representing them in $\mathcal{L}_H$ and $\mathcal{L}_K$.

Howson proved Theorem 7.32 in 1954. The topological techniques he used were quite different than the language-theoretic approach given above. In addition to knowing that $H \cap K$ is finitely generated it would be nice to have a method for bounding the rank of $H \cap K$. In his paper Howson established:

$$\mathrm{rk}(H \cap K) \le 2\mathrm{rk}(H)\mathrm{rk}(K) - \mathrm{rk}(H) - \mathrm{rk}(K) + 1.$$

In papers published a few years later Hanna Neumann improved Howson's bound, and conjectured that the rank of $H \cap K$ is at most

$$(\mathrm{rk}(H) - 1)(\mathrm{rk}(K) - 1) + 1.$$

This question is still open, and is referred to as the *Hanna Neumann Conjecture*. Hanna Neumann's son, Walter Neumann, has suggested a strengthened version of the conjecture [Ne90], and this strengthened version can be formulated as a purely graph theoretic conjecture [Di94].

**Remark 7.36.** The reader who would like to continue exploring connections between formal language theory and the study of infinite groups might begin by consulting Bob Gilman's survey article [Gi05].

### Exercises

(1)  Show that the language $\{a^k \mid k \equiv 0 \text{ or } 2 \pmod 5\}$ is a regular language.
(2)  Prove Lemma 7.9: If $\mathcal{L}$ is the language associated to a deterministic automaton, then $\mathcal{L}$ is the language associated to a complete, deterministic automaton.
(3)  Let $\mathcal{L}$ be a language. Define its reverse language to be

$$\mathrm{Rev}(L) = \{x_n \cdots x_1 \mid x_1 \cdots x_n \in \mathcal{L}\}.$$

Show that if $\mathcal{L}$ is a regular language then its reverse is a regular language.

(4) Let $\mathcal{L}$ be a regular language. Show that $\mathcal{L}$ is the language of a finite (but not necessarily deterministic) automaton with:

- a single start state (that may be an accept state as well)
- at most one more accept state
- no $\varepsilon$-transitions.

Further, show that the start state is not at the end of any transition and the accept state is not the initial vertex of any transition.

(5) Fix an alphabet $\mathcal{S}$ and a language $\mathcal{L} \subset \mathcal{S}^*$. For any two words $w, w' \in \mathcal{S}^*$ define a relation

$$w \sim w' \text{ if and only if } \mathrm{Cone}(w) = \mathrm{Cone}(w').$$

Prove that this is an equivalence relation on $\mathcal{S}^*$ and rephrase the Myhill–Nerode Theorem in terms of this equivalence relation.

(6) Determine if the following languages, contained in $\{a, b\}^*$, are regular.

    a. $L = \{a^i b^j \mid i \text{ is even and } j \text{ is odd}\}$
    b. $L = \{a^i \mid i \text{ is a Fibonnaci number}\}$
    c. $L = \{a^i b^j \mid j > i\}$
    d. $L$ consists of all words with more $b$'s than $a$'s.

(7) Show that the set $\{x^i y^{-i} \mid i \in \mathbb{Z} \setminus \{0\}\}$ is a basis for a free subgroup of $\mathbb{F}_2$.

(8) Show that every finitely generated abelian group admits a regular normal form.

(9) Show that $\mathrm{BS}(1, 2)$ admits a regular normal form.

(10) In Lemma 6.4 we showed that every element $g$ in the finitely generated, infinite torsion group, $\mathcal{B}$, can be expressed as:

$$g = \sigma^k x_1 x_2 \cdots x_n$$

where $k \in \{0, 1, 2\}$ and $x_1 x_2 \cdots x_n \in \{w, \sigma w \sigma^{-1}, \sigma^2 w \sigma^{-2}\}$. Why isn't this a regular normal form?

(11) Prove Proposition 7.26: If $G$ and $H$ have regular normal forms, then so does $G \oplus H$ and $G * H$.

(12) Prove that the group generated by the reflections in the sides of an equilateral triangle, $W_{333}$, admits a regular normal form.

(13) Let $G = \mathbb{F}_2 \oplus \mathbb{Z}$ with $\{x, y\}$ being the basis for $\mathbb{F}_2$ and $z$ a cyclic generator of $\mathbb{Z}$. Prove that the kernel $K$ of the map $\phi : \mathbb{F}_2 \oplus \mathbb{Z} \to \mathbb{Z}$, induced by $\phi(x) = \phi(y) = \phi(z) = 1$, is a free group generated by $\{x^{-1}z, y^{-1}z\}$.

(14) The even subgroup of $\mathbb{F}_2$ is a free group of rank 3. In Exercise 14 in Chapter 3 you showed that the kernel of $\phi : \mathbb{F}_2 \to \mathbb{Z}_3$ is a free group of rank 4. What is the rank of their intersection?

# 8

# The Lamplighter Group

Brief explanation: a lamplighter lives in a town whose street map is $C$ and walks about the town lighting various lamps (elements of $K$, especially when $K = \mathbb{Z}/2\mathbb{Z}$).

–Walter Parry

A short drive from Lafayette College (in Easton, Pennsylvania) one can find barns decorated in a classic Pennsylvania Dutch style with Hex diagrams placed around on the sides of the barn. In Figure 8.1 there is a simplified version of this, where copies of a single Hex diagram have been placed at the corners of a square.

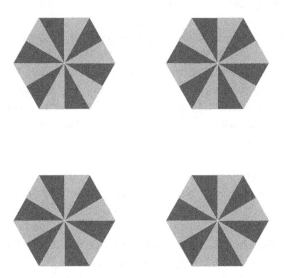

Fig. 8.1. An arrangement of simple Hex diagrams.

151

Let $G$ be the color-preserving symmetry group of this pattern. If we want to understand $G$ – perhaps at the level of being able to draw the Cayley graph of $G$ – we need some basic information about its structure. How many elements are there in $G$? Is there a natural generating set? Given that set of generators, is there a normal form for the elements of $G$?

The Hex designs only exhibit rotational symmetry, so their individual symmetry groups are cyclic of order 6. Because the Hex designs are disjoint, the subgroup of $G$ consisting of individual rotations with no permuting of the locations of the Hexes is isomorphic to the direct product $(\mathbb{Z}_6)^4 = \mathbb{Z}_6 \oplus \mathbb{Z}_6 \oplus \mathbb{Z}_6 \oplus \mathbb{Z}_6$. This gives a total of $6^4 = 1296$ symmetries, but this does not account for all of the symmetries of the Hex arrangement. Because the four Hex designs are identical, we can freely permute them. Thus we also have $\text{SYM}_4$ showing up as a subgroup of the symmetry group of this Hex arrangement.

On the other hand, any symmetry of the Hex arrangement can be described by permuting the Hex designs (without rotation), and then rotating each individual design. Thus each symmetry can be described by a pair consisting of an element from $(\mathbb{Z}_6)^4$ and a permutation from $\text{SYM}_4$. It is easy to see that this description is unique, hence the order of $G$ is $6^4 \cdot 4! = 31104$. This is certainly too large for us to draw the complete Cayley graph, but our analysis so far is quite helpful in giving a general description of what the Cayley graph looks like.

**Exercise 8.1.** If $A$ and $B$ are normal subgroups of $G$, where $G = A \cdot B$ and $A \cap B = e \in G$, then $G \approx A \oplus B$. We have just argued that there are two subgroups of the symmetry group of our Hex diagram, $(\mathbb{Z}_6)^4$ and $\text{SYM}_4$, whose product is $G$. Further, the intersection of these subgroups is the identity. Prove that $G$ is not the direct product of these two subgroups, by showing that the $\text{SYM}_4$ subgroup is not a normal subgroup of $G$.

The exercise above shows that $G$ is not the direct product of $(\mathbb{Z}_6)^4$ and $\text{SYM}_4$; $G$ is, however, a "semi-direct product" of $(\mathbb{Z}_6)^4$ and $\text{SYM}_4$, denoted $(\mathbb{Z}_6)^4 \rtimes \text{SYM}_4$. To see the distinction between direct and semi-direct products, we should examine how the binary operation works. In order to emphasize that we are not working with a direct product, we denote elements of $G$ using square brackets:

$$G = \{[(a, b, c, d), \sigma] \mid (a, b, c, d) \in (\mathbb{Z}_6)^4 \text{ and } \sigma \in \text{SYM}_4\}.$$

Now consider the product

$$[(1, 0, 0, 4), (34)] \cdot [(0, 0, 2, 1), (23)].$$

The corresponding action on the Hex arrangement can then be expressed in four steps:

1. The second Hex and third Hex are exchanged.
2. The Hex in the third position is rotated through an angle of 120° and the Hex in the fourth position is rotated through an angle of 60°.
3. The third Hex and fourth Hex are exchanged.
4. The first Hex is rotated 60° and the fourth Hex is rotated 240°.

So, in total, what happened?

1. The first Hex is never permuted, and it is rotated through an angle of 60°.
2. The second Hex is sent to the third position, then it is rotated 120°; after this it is moved into the fourth position and rotated an additional 240°. Thus at the end of this process, the second Hex ends up in the fourth position, where it is unrotated.
3. A similar analysis shows the third Hex ends up in the second position, unrotated.
4. The fourth Hex ends up in the third position, rotated 60°.

Thus we have:

$$[(1, 0, 0, 4), (34)] \cdot [(0, 0, 2, 1), (23)] = [(1, 0, 1, 0), (243)].$$

Notice that the permutations were unaffected by the presence of the subgroup of individual rotations, but the rotations seem to have been affected by the permutations. In fact, we can expand the equation above to make this explicit:

$$[(1, 0, 0, 4), (34)] \cdot [(0, 0, 2, 1), (23)]$$
$$= [(1, 0, 0, 4) + (0, 0, \mathbf{1}, \mathbf{2}), (34)(23)]$$
$$= [(1, 0, 1, 0), (243)].$$

The key insight is that, in moving the permutation (34) past the rotations $(0, 0, 2, 1)$, we actually apply the permutation to the 4-tuple of rotations. This example, we hope, provides some insight into and motivation for the following construction.

**Definition 8.2.** Let $H$ be a group and let $\phi : K \to \text{Aut}(H)$ be a group homomorphism from $K$ to the automorphism group of $H$. For notational convenience below, let $\phi_k$ denote the automorphism $\phi(k) \in \text{Aut}(H)$. The elements of the associated *semi-direct product* are ordered pairs of elements $[h, k]$. The binary operation is:

$$[h_1, k_1] \cdot [h_2, k_2] = [h_1 \phi_{k_1}(h_2), k_1 k_2].$$

Since $\phi_{k_1}(h_2) \in H$, the product $h_1 \phi_{k_1}(h_2)$ is computed in $H$; the product $k_1 k_2$ is computed in $K$.

It is a not-particularly-pleasant exercise to verify that this set of elements and binary operation does indeed define a group. It is denoted $H \rtimes K$, or $H \rtimes_\phi K$ when one needs to be explicit about the homomorphism $\phi$ from $K$ to $\text{Aut}(H)$.

We leave the proof of the following result as an exercise.

**Theorem 8.3.** *Let $H$ be a group, let $\phi : K \to \text{Aut}(H)$, and let $G$ be the semi-direct product $H \rtimes K$.*

1. *The map from $H$ to $G$ given by $h \mapsto [h, e_K]$ is an injective homomorphism.*
2. *The image of $H$ in $G$ (as above) is a normal subgroup.*
3. *The map from $K$ to $G$ given by $k \mapsto [e_H, k]$ is an injective homomorphism.*
4. *The quotient of $G$ mod $H$ is isomorphic to $K$, $G/H \approx K$.*
5. *There is a conjugation relation that in shorthand says*

$$khk^{-1} = \phi_k(h),$$

*or more accurately:*

$$[e_H, k] \cdot [h, e_K] \cdot [e_H, k^{-1}] = [\phi_k(h), e_K].$$

**Example 8.4.** The most commonly encountered examples of semi-direct products are the dihedral groups. Here we have $D_n \approx \mathbb{Z}_n \rtimes_\phi \mathbb{Z}_2$. The subgroup $\mathbb{Z}_n$ consists of the rotations and the indicated homomorphism $\phi : \mathbb{Z}_2 \to \text{Aut}(\mathbb{Z}_n)$ takes the non-identity element of $\mathbb{Z}_2$ to the automorphism of $\mathbb{Z}_n$ given by sending each element to its inverse. Item 5 of Theorem 8.3 is essentially just the familiar fact that $f\rho^i f = \rho^{-i}$, where $f$ is a chosen reflection, and $\rho$ is a rotation.

**Example 8.5.** Let $\alpha$ be the automorphism of $\mathbb{Z}_7$ given by $\alpha(x) = 2x$, so that $\alpha^3(x) = 2^3 x \equiv x \pmod 7$. There is then a homomorphism $\phi : \mathbb{Z}_3 \to \text{Aut}(\mathbb{Z}_7)$ given by $\phi(1) = \alpha$. The induced semi-direct product

$\mathbb{Z}_7 \rtimes \mathbb{Z}_3$ is a non-abelian group of order 21. In fact, it is the unique non-abelian group of order 21. The Cayley graph, with respect to $\{a, b\}$ where $b = 1 \in \mathbb{Z}_7$, is shown in Figure 8.2.

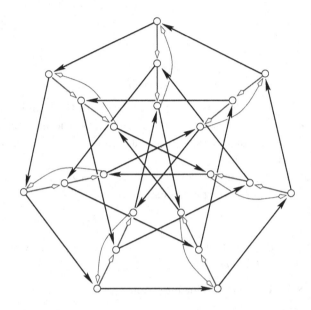

Fig. 8.2. A Cayley graph of the non-abelian group of order 21.

Similarly, let $\phi$ be the automorphism of $\mathbb{Z}_9$ given by $\phi(x) = 4x$, so that $\phi$ is of order 3 in $\mathrm{Aut}(\mathbb{Z}_9)$. Then we can form a non-abelian group of order 27, $\mathbb{Z}_9 \rtimes \mathbb{Z}_3$. The Doyle graph, shown in Fig 1.29, is the underlying graph of a Cayley graph of this group. If we let $a$ be an element generating $\mathbb{Z}_9$ and $b$ an element generating $\mathbb{Z}_3$, then the Doyle graph is the Cayley graph for $\mathbb{Z}_9 \rtimes \mathbb{Z}_3$ with respect to the generating set $\{a, c = ba^{-1}\}$.

**Example 8.6.** The group $\mathbb{Z}^2 = \mathbb{Z} \oplus \mathbb{Z}$ admits an automorphism $\alpha$ of order 2, given by $\alpha\left[(a, b)\right] = (b, a)$. Thus we may form a semi-direct product $\mathbb{Z}^2 \rtimes \mathbb{Z}_2$. We will return to this seemingly simple example in Section 9.4.

**Example 8.7.** The semi-direct product $H \rtimes \mathrm{Aut}(H)$ is called the *holomorph* of $H$. In some vague sense, best not to be made precise, the holomorph of $H$ is the largest meaningful semi-direct product one can make starting from $H$. As an example, we note that $\mathrm{Hol}(\mathbb{Z}_2 \oplus \mathbb{Z}_2) \approx \mathrm{SYM}_4$. (This isomorphism is not obvious.)

The "wreath product" of two groups $G$ and $H$ is defined using semi-direct products. One starts by forming a direct sum of copies of $G$, one copy for each $h \in H$. In fact, one often indexes the copies of $G$ by the elements of $H$, hence this direct sum can be expressed as $\bigoplus_{h \in H} G$. The *wreath product* – denoted $G \wr H$ – is the semi-direct product of $\bigoplus_{h \in H} G$ and $H$:

$$G \wr H = \left( \bigoplus_{h \in H} G \right) \rtimes H.$$

The action of $H$ on the direct sum defined is via the standard Cayley action of $H$ on itself. Thus, given $\vec{g} \in \bigoplus_{h \in H} G$, the element $h \in H$ permutes the entries of $\vec{g}$ by taking the entry in position $h'$ to the position $h \cdot h'$ (for every $h' \in H$).

As an elementary example, consider $\mathbb{Z}_2 \wr \mathbb{Z}_3$. This group is a semi-direct product $(\mathbb{Z}_2)^3 \rtimes \mathbb{Z}_3$ where the associated automorphism

$$\phi : \mathbb{Z}_3 \to \mathrm{Aut}(\mathbb{Z}_2 \oplus \mathbb{Z}_2 \oplus \mathbb{Z}_2)$$

is given by declaring that $\phi(1) \in \mathrm{Aut}(\mathbb{Z}_2 \oplus \mathbb{Z}_2 \oplus \mathbb{Z}_2)$ is the cyclic permutation $\alpha(a, b, c) = (b, c, a)$.

**Exercise 8.8.** Verify the following equation:

$$[(0, 1, 1), 1] \cdot [(1, 0, 0), 1] = [(0, 1, 0), 2] \in \mathbb{Z}_2 \wr \mathbb{Z}_3$$

The remainder of this chapter is devoted to understanding a single example of this wreath product construction, $\mathbb{Z}_2 \wr \mathbb{Z}$. This group is a semi-direct product:

$$\mathbb{Z}_2 \wr \mathbb{Z} = \left( \bigoplus_{i \in \mathbb{Z}} \mathbb{Z}_2 \right) \rtimes \mathbb{Z} = (\cdots \oplus \mathbb{Z}_2 \oplus \mathbb{Z}_2 \oplus \mathbb{Z}_2 \oplus \cdots) \rtimes \mathbb{Z}.$$

The infinite direct sum has as its elements $\mathbb{Z}$-tuples of elements of $\mathbb{Z}_2$, only finitely many of which are allowed to be non-zero.[1]

This particular wreath product of groups is called the *lamplighter group*, for reasons that will become clear, and it is denoted $L_2$. For convenience, call the subgroup of $L_2$ given by $\bigoplus_{i \in \mathbb{Z}} \mathbb{Z}_2$ "$A$." Since an element of $A$ is determined by which (finite number) entries have non-zero elements, we can represent the elements of $A$ as finite subsets of

---

[1] The reader may have encountered the phrases "direct product" and "direct sum" in various contexts and might believe they are synonymous. They are not. In a direct sum, often denoted $\oplus$, only finitely many entries are allowed to be non-identity elements; in a direct product, often denoted $\times$, there is no restriction on the number of non-identity elements.

integers. The empty set would then correspond to the identity element, $\overline{0} = (\ldots, 0, 0, 0, \ldots) \in A$. The binary operation in $A$ is coordinate-wise addition in $\mathbb{Z}_2$. Viewing elements of $A$ as subsets of $\mathbb{Z}$, we see that this corresponds to the symmetric difference

$$S \triangle T = \{n \in \mathbb{Z} \mid n \in S \text{ or } n \in T \text{ but } n \notin S \cap T\}.$$

In particular, $\{-2, 0, 1\} \triangle \{-3, 0, 4\} = \{-3, -2, 1, 4\}$. Thus every element in $L_2$ can be represented by an ordered pair $[S, n]$ where $S$ is a subset of integers, and $n \in \mathbb{Z}$. The binary operation is given by

$$[S, n] \cdot [T, m] = [S \triangle (T + n), n + m]$$

where $T + n = \{t + n \mid t \in T\}$. The identity element is $[\emptyset, 0]$.

As the subgroup $A < L_2$ is an infinite direct sum, the following lemma may at first seem a bit surprising.

**Lemma 8.9.** *The lamplighter group $L_2$ can be generated by two elements: one of order 2 and the other of infinite order.*

*Proof.* Let $t$ be the element corresponding to $[\emptyset, 1] \in L_2$ and let $a$ be the element $[\{0\}, 0] \in L_2$. The action of $\mathbb{Z}$ on $A$ can be described by noticing that $\phi(1) \in \text{Aut}(A)$ has the effect of adding 1 to each coordinate. Thus

$$ta = [\emptyset, 1] \cdot [\{0\}, 0] = [\emptyset \triangle \{1\}, 1 + 0] = [\{1\}, 1]$$

and more generally

$$t^n a = [\emptyset, n] \cdot [\{0\}, 0] = [\{n\}, n]$$

for any $n \in \mathbb{Z}$. Thus

$$t^n a t^{-n} = [\{n\}, 0].$$

It follows that an arbitrary element of the subgroup $A$ can be expressed as:

$$[\{n_1, n_2, \ldots, n_m\}, 0] = t^{n_1} a t^{-n_1} \cdot t^{n_2} a t^{-n_2} \cdots t^{n_m} a t^{-n_m}.$$

Further, an arbitrary element of $L_2$ can be expressed as:

$$[\{n_1, n_2, \ldots, n_m\}, k] = t^{n_1} a t^{-n_1} \cdot t^{n_2} a t^{-n_2} \cdots t^{n_m} a t^{-n_m} \cdot t^k.$$

Therefore, the set $\{a, t\}$ is a generating set for the lamplighter group. $\square$

The perspective of this book is geometric, and it is this perspective that illuminates $L_2$ and explains why it is called the lamplighter group. We know that an arbitrary element of $L_2$ can be expressed as $[\{n_1, n_2, \ldots, n_m\}, k]$. To think of this geometrically, take the Cayley

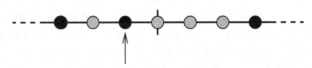

Fig. 8.3. A geometric representation of an element of the lamplighter group. The central vertex corresponds to the identity element in $\mathbb{Z}$, so this element of $L_2$ is $[\{-2, 0, 1, 2\}, -1]$.

graph of $\mathbb{Z}$ and color the vertices corresponding to $\{n_1, n_2, \ldots, n_m\}$ yellow; color all of the other vertices black; and add a pointer pointing to the vertex associated to $k$. (An example is shown in Figure 8.3.) We will refer to such a decorated graph as the *picture* of an element. The picture of the identity element consists of all black vertices with the pointer pointing at zero.

If we want to study the Cayley graph with respect to the generating set $\{a, t\}$, then we need to determine the effect of right multiplying by an $a$ or $t$. If $g = [S, k]$ is an arbitrary element of $L_2$, then

$$g \cdot a = [S, k] \cdot a = [S, k] \cdot [\{0\}, 0] = [S \triangle \{k\}, k + 0] = [\widehat{S}, k]$$

where $\widehat{S}$ either adds $k$ to $S$ (if the vertex at position $k$ was colored black) or it removes $k$ from $S$ (if the vertex at position $k$ was colored yellow). Thus, in terms of the pictures of elements, right multiplying an arbitrary element by $a$ has the effect of changing the color of the vertex that the pointer is pointing at. Right multiplication by $t$ gives:

$$g \cdot t = [S, k] \cdot t = [S, k] \cdot [\emptyset, 1] = [S, k + 1]$$

which simply moves the pointer one unit to the right. Similarly, right multiplication by $t^{-1}$ moves the pointer one unit to the left.

We are now in a position to explain the intuition behind these pictures of the elements of $L_2$. We have "lit lamps" at all the positions where there are non-zero entries in the infinite direct sum of copies of $\mathbb{Z}_2$, all other lamps are "unlit," and our lamplighter is stationed at position $k$. Right multiplication by $a$ has the effect of lighting or dowsing the lamp that the lamplighter is stationed at; right multiplication by $t$ or $t^{-1}$ moves the lamplighter to the right or left. (See Walter Parry's quotation at the beginning of this chapter.)

The use of these pictures makes it simple to express an arbitrary element in terms of the generators. Take, for instance the element of $L_2$ shown in Figure 8.4. To create the element corresponding to the picture in Figure 8.4, starting from the identity, we could:

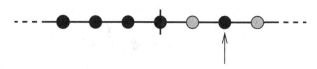

Fig. 8.4. An element of $L_2$, which can be expressed as $t^3 a t^{-2} a t$ or $t a t^2 a t^{-1}$.

1. Move the lamplighter three units to the right.
2. Light the lamp at $x = 3$.
3. Move the lamplighter two units to the left.
4. Light the lamp at $x = 1$.
5. Move the lamplighter one unit to the right.

This procedure shows that this element can be expressed as $t^3 a t^{-2} a t$. Of course there are other ways one could express this particular element. For example, one could follow a slightly different process:

1. Move the lamplighter one unit to the right.
2. Light the lamp at $x = 1$.
3. Move the lamplighter two units to the right.
4. Light the lamp at $x = 3$.
5. Move the lamplighter one unit to the left.

From this we see that the same element can be expressed as $t a t^2 a t^{-1}$. In other words, $t^3 a t^{-2} a t = t a t^2 a t^{-1}$ in $L_2$.

In Figure 8.5 we show a cycle in the Cayley graph of $L_2$ (with respect to $\{a, t\}$) that corresponds to the relation $a t a t^{-1} = t a t^{-1} a$. We have denoted the vertex corresponding to $g \in L_2$ by its picture.

## Exercises

(1)  Prove Theorem 8.3.
(2)  Prove that the symmetry group of a cube (Section 1.6.3) is isomorphic to $(\mathbb{Z}_2 \oplus \mathbb{Z}_2 \oplus \mathbb{Z}_2) \rtimes \text{SYM}_3$.
(3)  A finite subgroup of $\text{Aut}(\mathbb{F}_n)$, $\Omega_n$, was introduced in Section 3.10. Prove that $\Omega_n \approx \underbrace{(\mathbb{Z}_2 \oplus \cdots \oplus \mathbb{Z}_2)}_{n \text{ copies}} \rtimes \text{SYM}_n$. (What does this have to do with Exercise 2?)
(4)  Prove that $\text{SYM}_n \approx A_n \rtimes \mathbb{Z}_2$.
(5)  Prove $\text{Hol}(\mathbb{Z}_2 \oplus \mathbb{Z}_2) \approx \text{SYM}_4$.
(6)  Verify that Figure 8.2 is correct.

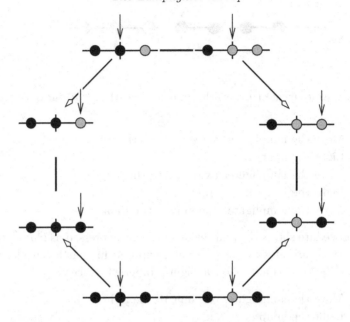

Fig. 8.5. A cycle in the Cayley graph of the lamplighter group, where undirected edges correspond to the generator $a$ and directed edges to the generator $t$.

(7) There are five distinct groups of order 12. You have probably encountered four of them: $\mathbb{Z}_{12} \approx \mathbb{Z}_3 \oplus \mathbb{Z}_4$, $\mathbb{Z}_6 \oplus \mathbb{Z}_2$, $D_6 \approx D_3 \oplus \mathbb{Z}_2$, and $A_4$. The fifth group of order 12 can be described using semi-direct products. Let $H = \mathbb{Z}_3$ and let $K = \mathbb{Z}_4$, where $H$ is generated by $h$ and $K$ is generated by $k$. Let $\phi : K \to \mathrm{Aut}(\mathbb{Z}_3)$ by $\phi(k) = \alpha$ where $\alpha(h) = h^2$. Show that $G = H \rtimes_\phi K$ is a group of order 12 that is not isomorphic to any of the four previously listed groups of order 12.

(8) Prove that $D_4 \approx \mathbb{Z}_2 \wr \mathbb{Z}_2$.

(9) Let $G$ and $H$ be two finite groups. Show that the order of $G \wr H$ is $|G|^{|H|} \cdot |H|$.

(10) In this exercise we examine the symmetries of rooted binary trees. (See Chapter 6.)

    a. Let $T$ denote the finite, rooted binary tree with eight leaves shown in Figure 8.6. Prove that $\mathrm{Sym}(T) \approx (\mathbb{Z}_2 \wr \mathbb{Z}_2) \wr \mathbb{Z}_2$.

    b. Let $\mathcal{T}$ denote the infinite, rooted binary tree where every vertex has two descendants. Prove that $\mathrm{Sym}(\mathcal{T}) \approx \mathrm{Sym}(\mathcal{T}) \wr \mathbb{Z}_2$.

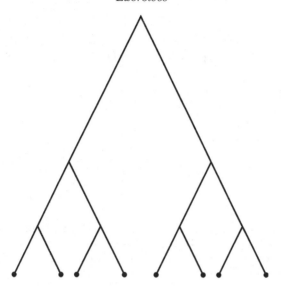

Fig. 8.6. A finite, rooted binary tree.

(11) Prove that the wreath product of two finitely generated groups is finitely generated.

(12) Show that $at^n at^{-n} = t^n at^{-n} a \in L_2$, for any $n \in \mathbb{N}$.

(13) The Cayley graph $\Gamma$ of $L_2$ with respect to $\{a, t\}$ is fairly complicated. You can gain some intuition for this by drawing the ball of radius 3, $\mathcal{B}(e, 3) \subset \Gamma$. You should find that this is a tree, but you should also realize that $\mathcal{B}(e, 4)$ is not a tree, as is shown by Figure 8.5.

# 9

# The Geometry of Infinite Groups

[E]ssentially all geometric constructs which are global in nature, such as paths of shortest length, global manifestations of curvature, planes, half-spaces, rates of growth, which are studied in differential geometry have manifestations in combinatorial approximations to that geometry.

–James Cannon

## 9.1 Gromov's Corollary, aka the Word Metric

We have seen a number of examples of groups acting on the real line. Sometimes these actions preserve the distance between points on the line, such as the action of $D_\infty$. Other actions we have considered do not preserve distances, for example, the action $\mathrm{BS}(1,2) \curvearrowright \mathbb{R}$ presented in Chapter 4. One of the most powerful insights in the study of finitely generated infinite groups is that they can always be viewed as groups acting in a distance-preserving way on a geometric object. We refer to this insight as *Gromov's Corollary* to Cayley's Better Theorem, in honor of a groundbreaking paper that Mikhail Gromov wrote in the 1980s [Gr87], which highlighted this perspective, introduced a number of questions motivated by the geometry of infinite groups, and introduced powerful tools that can be used to answer them.

In order to present Gromov's Corollary, we need to introduce a reasonably flexible notion of "geometric object" as well as formally define what we mean by saying that a group action "preserves distances."

**Definition 9.1.** A *metric space* consists of a set $X$ and a distance function $d : X \times X \to \mathbb{R}$ such that, for any $x, y$, and $z \in X$:

1. $d(x, y) \geq 0$,
2. $d(x, y) = 0$ if and only if $x = y$,

3. $d(x, y) = d(y, x)$,
4. $d(x, y) + d(y, z) \geq d(x, z)$.

This final condition is referred to as the *triangle inequality*. Examples of metric spaces include the real line as well as $\mathbb{R}^n$ with their usual notion of distance. It also includes objects such as spheres, where the distance between two points is declared to be the length of the shortest path connecting the two points.

A function from one metric space to another, $\phi : X_1 \to X_2$, is an *isometry* if it is onto and, for all $x, y \in X$, $d_1(x, y) = d_2(\phi(x), \phi(y))$. (In other words, the function preserves ("iso") distances ("metry").) If $G \curvearrowright X$, then the action is *by isometries*, or is an *isometric action*, if for all $x, y \in X$ and $g \in G$, one has:

$$d(x, y) = d(g \cdot x, g \cdot y).$$

The three reflection groups $W_{244}, W_{236}$ and $W_{333}$ actions on $\mathbb{R}^2$ (discussed in Chapter 2) are isometric actions.

**Theorem 9.2** (Gromov's Corollary). *Every finitely generated group can be faithfully represented as a group of isometries of a metric space.*

Similar to the proofs of Cayley's Theorem and Cayley's Better Theorem, the metric space is built from the group $G$. Because the vertices of a Cayley graph correspond to group elements, one can use the distance between vertices on a Cayley graph to define a distance function on the set of elements of the group. Further, one can phrase this entirely in terms of group elements, with no direct mention of the Cayley graph. If $S$ is a finite set of generators for $G$, and $g, h \in G$, then set

$$d_S(g, h) = \text{the length of the shortest word representing } g^{-1}h \ .$$

If $\omega$ is a word in $\{S \cup S^{-1}\}^*$ representing $g^{-1}h$ then

$$g^{-1}h = \pi(\omega) \Rightarrow h = g\pi(\omega).$$

So $\omega$ labels a path connecting the vertex associated to $g$ to the vertex associated to $h$. Thus a minimal-length word describes a minimal-length path between the associated vertices in the Cayley graph $\Gamma_{G,S}$.

If $S$ is a generating set for $G$, then we have previously discussed the length of words in $\{S \cup S^{-1}\}^*$. We can extend this to the elements of $G$.

**Definition 9.3.** Let $G$ be a group with a fixed, finite generating set $S$. Then the *length* of $g \in G$ is the minimal-length of a word $\omega \in \{S \cup S^{-1}\}^*$

where $\pi(\omega) = g$. Denote this value by $|g|$. Notice, in particular, that $|g|$ represents the minimal length of a path from the vertex representing $e$ to the vertex representing $g$ in the Cayley graph $\Gamma_{G,S}$.

As a simple example, if $G = \mathbb{Z}\oplus\mathbb{Z}$, $S = \{(1,0),(0,1)\}$ is the generating set, and $g = (m,n) \in \mathbb{Z} \oplus \mathbb{Z}$, then the length of $G$ is just $|m| + |n|$.

This metric on the group $G$ is often referred to as the *word metric* and it does depend on the choice of generators. This is the reason that $S$ appears as a subscript on the distance function. We reserve the right to be sloppy and drop this subscript from time to time, when the fixed set of generators is either clear or not of central importance to the argument.

*Proof of Gromov's Corollary.* Conditions 1 and 2 have already been established, so it only remains to show that the distance function is symmetric and that the triangle inequality holds.

The function $d_S$ is symmetric: if $d_S(g,h) = n$ then there is a word $\omega$ of length $n$ such that $g^{-1}h = \pi(\omega)$. But then $h^{-1}g = \pi(\omega^{-1})$ and so $d_S(h,g)$ is at most $n$. But if there were a shorter word representing $h^{-1}g$, then we could take its formal inverse and form a shorter word representing $g^{-1}h$. Thus $d_S(h,g) = d_S(g,h)$ for any $g,h \in G$.

Our function $d_S$ also satisfies the triangle inequality:

$$d_S(g,k) + d_S(k,h) \geq d_S(g,h)$$

for all $g,h$ and $k \in G$. To see this, let $\omega_{gk}$ and $\omega_{kh}$ be minimal-length words such that $g \cdot \omega_{gk} = k$ and $k \cdot \omega_{kh} = h$. Thus $g \cdot \pi(\omega_{gk}\omega_{kh}) = h$, hence

$$d_S(g,h) \leq |\omega_{gk}| + |\omega_{kh}| = d_S(g,k) + d_S(k,h).$$

The comments above show that the group $G$ can be viewed as a metric space. Cayley's Theorem shows that left multiplication gives an action of $G$ on itself, and this action preserves distances. That is,

$$d_S(h,k) = |h^{-1}k| = |h^{-1}g^{-1}gk| = d_S(gh,gk)$$

for any $g,h,k \in G$. □

**Example 9.4.** A standard Cayley graph of $\mathbb{Z}\oplus\mathbb{Z}$ is shown in Figure 9.1. The distance between the lower left-hand vertex and the upper right-hand vertex is 7. This distance is realized by a total of $\binom{7}{4}$ different words of length 7. To see this, let the generator corresponding to horizontal arrows be $x$ and the generator corresponding to vertical arrows be $y$. Then any word corresponding to a minimal-length path between

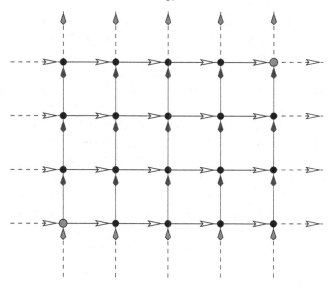

Fig. 9.1. The distance between the lower left-hand vertex and the upper right-hand vertex is 7.

these vertices consists of seven letters: four $x$'s and three $y$'s. Such a word is determined by the placement of the $x$'s (or by the placement of the $y$'s). For example, the edge path going across the bottom of the figure and then the right edge corresponds to $xxxxyyy$ while a diagonal path is given by $xyxyxyx$. In fact, the union of all edges that occur in minimal-length paths between these two vertices forms the rectangle that is the focus of this figure.

**Definition 9.5.** A minimal-length edge path joining two vertices $v$ and $w$ is a *geodesic path*. Similarly, a word $\omega$ is a *geodesic*[1] if $|\omega| = |\pi(\omega)|$. In other words, a word is a geodesic if the corresponding path in the Cayley graph is a minimal-length path between its endpoints. In the example above we have essentially pointed out there are $\binom{7}{4}$ geodesic words expressing the group element $(4, 3) \in \mathbb{Z} \oplus \mathbb{Z}$.

**Remark 9.6.** The Word Metric is primarily studied in the context of infinite groups, but it does show up from time to time in the context of finite groups. For example, the *diameter* of any finite graph $\Gamma$ is the minimal integer $D$ such that one can get between any two vertices in $\Gamma$

---

[1] In geometry, a path is a geodesic if it is (locally) of minimal-length. Examples include line segments in the Euclidean plane and arcs of great circles on spheres.

by an edge path of length $\leq D$. If $\Gamma$ happens to be a Cayley graph of a group $G$, the diameter represents the maximum length that is necessary to express any element in $G$ in terms of the chosen generators.

There is a famous example where the diameter of a Cayley graph is of popular interest. Let $\mathfrak{R}$ be the Rubik's Cube. The set of all symmetries of $\mathfrak{R}$ forms a very large, but still finite, group. The standard move in working with a Rubik's Cube is to rotate one of the six faces by 90°. These standard moves generate all possible moves, and so they form a set of generators for the Rubik's Cube group. The diameter of the Cayley graph of Sym($\mathfrak{R}$) with respect to this generating set gives the maximum number of steps necessary to get from an arbitrary position to the solution position using the standard moves. Since the Rubik's Cube group has over 43 quintillion elements, it is not a surprise that this diameter is not yet known. (Our most recent Google search shows that the diameter is known to be $\leq 26$.)

Finding geodesic words is not always as easy as it is in $\mathbb{Z} \oplus \mathbb{Z}$. The lamplighter group $L_2$ gives us a more illuminating example of some of the difficulties one can encounter. Before establishing a general formula for the length of an arbitrary element, we should consider the element $g \in L_2$ corresponding the picture in Figure 8.3. In order to construct such a configuration of lit lamps and pointer, the lamplighter must light the four indicated lamps, and she must end her travels at $x = -1$. The most efficient way to do this is to have the lamplighter travel to $x = 2$, then move back to $x = -2$, and finish at $x = -1$, remembering to turn on the appropriate lamps when in position. Thus any minimal-length expression for $g$ must involve $2 + 4 + 1 + 4 = 11$ letters from $\{a, t, t^{-1}\}$. This discussion indicates why we need to specify the following two constants associated to an arbitrary element $g = [S, k] \in L_2$.

$$R = \text{Max}\{S \cup \{0\}\},$$
$$L = \min\{S \cup \{0\}\}.$$

**Proposition 9.7.** *Let* $g = [S, k] \in L_2$, *and let* $R$ *and* $L$ *be as above. Then the length of* $g$ *is given by*

$$|g| = |S| + \min\{2R + |L| + |k - L|, 2|L| + R + |k - R|\}.$$

In order to establish this formula, we introduce two normal forms for $L_2$.

**The right-first normal form:** Every element of $L_2$ can be uniquely expressed by first moving the lamplighter to the position of the rightmost lit lamp, then moving to the leftmost lit lamp, lighting lamps each time the lamplighter is first in that position. Finally, move the lamplighter left or right, if necessary, to get the lamplighter into position indicated by the pointer.

As an example, consider the element $g = [\{-2, 0, 1, 2\}, -1] \in L_2$, which is shown in Figure 8.3. The right-first normal form expression for this element is:

$$g = atatat^{-4}at.$$

**The left-first normal form:** This is identical to the normal form above, except that one first moves to the left, and so on. If we again consider $g = [\{-2, 0, 1, 2\}, -1] \in L_2$ we see that the left-first normal form expression for this element is:

$$g = at^{-2}at^3 atat^{-3}.$$

Since we already know $|g| = 11$, it follows that the right-first normal form for $g$ is a geodesic word, while the left-first normal form for $g$ is not.

The descriptions above are not completely clear when applied to elements like $g = [\{1, 3\}, 2] \in L_2$, which is shown in Figure 8.4. The picture for this element contains no lit lamps to the left of zero. The right-first normal form for such an element simply moves the lamplighter right, lighting lamps along the way, and then moves the pointer to the correct ending position. The left-first normal form is the exact same.

**Exercise 9.8.** Find the right-first and left-first normal forms of the element $g = [\{-3, 1, 2\}, 4] \in L_2$.

*Proof of Proposition 9.7.* We first establish connections between the right-first and left-first normal forms for $g \in L_2$ and the expressions in the statement of the proposition.

Case 1: Assume there is a lit lamp at some $m < 0$, in which case $L < 0$. The right-first normal form will have $|S|$ occurrences of $a$ (for lighting the lamps). There will also be $2R + |L| + |k - L|$ occurrences of $t$ or $t^{-1}$, since the lamplighter will move from $0$ to $R$, then back to $L$ and from there to $k$ (the location of the pointer).

Case 2: Assume there are no lit lamps at negative integers, so $L = 0$. The right-first normal form will again have $|S|$ occurrences of $a$. The lamplighter will need to take $R$ steps to the right, and then will need to move $|k - R|$ steps to end at the location of the pointer. So the length of the right-first normal form is

$$|S| + R + |k - R| = |S| + 2|L| + R + |k - R|$$

since $L = 0$.

Case 3: Assume there is a lit lamp at some $n > 0$, in which case $R > 0$. Then, as in Case 1, the length of the left-first normal form is $|S| + 2|L| + R + |k - R|$.

Case 4: Assume there is no lamp lit at a positive integer, so $R = 0$. Then, as in Case 2, the length of the left-first normal form is

$$|S| + |L| + |k - L| = |S| + 2R + |L| + |k - L|.$$

Thus, by considering the right-first and left-first normal forms of $g$ we see that

$$|g| \leq |S| + \min\{2R + |L| + |k - L|, 2|L| + R + |k - R|\}.$$

It remains to be shown that this is actually an equality (which will imply that for any $g \in L_2$, either its left-first or right-first normal form is a geodesic). But if $g = [S, k]$ then, in any expression for $g$, there must be $|S|$ occurrences of the generator $a$. Further, the lamplighter must travel from 0 to every lit lamp, and then move to the final position of the pointer. By considering cases similar to the ones above, it can be shown that $\min\{2R + |L| + |k - L|, 2|L| + R + |k - R|\}$ is the minimal number of steps that needed. As each step the lamplighter takes requires an application of either $t$ or $t^{-1}$, we get

$$|g| \geq |S| + \min\{2R + |L| + |k - L|, 2|L| + R + |k - R|\}.$$

Hence the equality stated in the theorem holds.  □

## 9.2 The Growth of Groups, I

In Section 5.3 we introduced the idea of spheres and balls of radius $n$ in Cayley graphs. For a fixed generating set $S$, the associated distance function on the group $G$ is the same as the edge-path distance between vertices in the Cayley graph $\Gamma_{G,S}$. Thus the following re-definition is consistent with the earlier introduction of spheres.

**Definition 9.9.** Given a group element $g \in G$ and a positive integer $n$, the *sphere of radius $n$* is

$$S(g, n) = \{h \in G \mid d_S(g, h) = n\}.$$

The *ball of radius n* is

$$\mathcal{B}(g, n) = \{h \in G \mid d_S(g, h) \leq n\}.$$

Notice that there is a difference between balls of radius $n$ in a Cayley graph and balls of radius $n$ in the group. The ball of radius $n$, based at the vertex $v_g$ associated to $g$ in the Cayley graph $\Gamma_{G,S}$, has vertex set $\{v_h \mid h \in \mathcal{B}(g, n)\}$. However, the ball of radius $n$ in the Cayley graph also contains edges connecting these vertices to $v_g$.

The size of spheres and balls of radius $n$ in a Cayley graph are of considerable importance, as they give us a concrete way to talk about the growth of a given infinite group.

**Definition 9.10.** The *spherical growth function of a group $G$ with respect to a finite generating set $S$* is the function $\sigma : \mathbb{N} \to \mathbb{N}$ defined by $\sigma(n) = |\mathcal{S}(e, n)|$. The associated *growth series* is the formal power series

$$\mathcal{S}(z) = \sum_{n \geq 0} \sigma(n) z^n = \sum_{g \in G} z^{|g|}.$$

**Example 9.11.** Let $G = \mathbb{Z}$ with a single generator $t$. Then

$$\sigma(n) = \begin{cases} 1 & n = 0 \\ 2 & n > 0 \end{cases}$$

and so the associated growth series is

$$\mathcal{S}(z) = 1 + 2z + 2z^2 + 2z^3 + \cdots.$$

In studying formal power series, combinatorialists often ask if a given series is *rational*, meaning it is the power series associated to a rational function. In the situation above, it is elementary to exhibit a rational function corresponding to the growth series:

$$\mathcal{S}(z) = 1 + 2z + 2z^2 + 2z^3 + \cdots$$

$$= 2(1 + z + z^2 + z^3 + \cdots) - 1 = 2\frac{1}{1-z} - 1$$

$$= \frac{2}{1-z} - \frac{1-z}{1-z} = \frac{1+z}{1-z}.$$

If $\mathbb{Z}$ is generated by $\{2, 3\}$ then one would expect a fairly different growth function and associated series. (The Cayley graphs for these two generating sets are shown in Figure 1.20.) Let $\mathcal{S}_{\{2,3\}}(n)$ be the sphere of radius $n$, with respect to $\{2, 3\}$, and let $\sigma_{\{2,3\}}(n)$ be the size of $\mathcal{S}_{\{2,3\}}(n)$. By inspection we find that $\sigma_{\{2,3\}}(0) = 1$ and $\sigma_{\{2,3\}}(1) = 4$;

a short computation shows $\mathcal{S}_{\{2,3\}}(2) = \{\pm 1, \pm 4, \pm 5, \pm 6\}$ and therefore $\sigma_{\{2,3\}}(2) = 8$. After this, the pattern becomes simple. Because 2 and 3 are relatively prime, any integer can be expressed as $n = l \cdot 2 + m \cdot 3$ where $l$ and $m$ are integers. But is this a geodesic expression? That is, is $|l| + |m|$ as small as it can be in order to produce $n$? Not if $|l| \geq 3$, for then the substitution $3 \cdot \underline{2} = 2 \cdot \underline{3}$ would reduce the length (where our generators have been underlined). Thus a minimal-length word representing $n \in \mathbb{Z}$ contains at most two instances of $\underline{2}$. So the sphere of radius $n$ contains at most ten elements: $\pm n \cdot \underline{3}$, $\pm(n-1) \cdot \underline{3} \pm \underline{2}$ and $\pm(n-2) \cdot \underline{3} \pm 2 \cdot \underline{2}$. However, for $n \geq 3$, $(n-1) \cdot \underline{3} - \underline{2} = (n-3) \cdot \underline{3} + 2 \cdot \underline{2}$, so $(n-1) \cdot \underline{3} - \underline{2}$ is actually in $\mathcal{S}_{\{2,3\}}(n-1)$. Further, $(n-2) \cdot \underline{3} - 2 \cdot \underline{2} = (n-4) \cdot \underline{3} + \underline{2}$, so $(n-2) \cdot \underline{3} - 2 \cdot \underline{2} \in \mathcal{S}_{\{2,3\}}(n-1)$. So for $n \geq 3$, the sphere of radius $n$ is

$$\mathcal{S}_{\{2,3\}}(n) = \{\pm n \cdot \underline{3}, \pm[(n-1) \cdot \underline{3} + \underline{2}], \pm[(n-2) \cdot \underline{3} + 2 \cdot \underline{2}]\}$$

and therefore

$$\sigma_{\{2,3\}}(n) = \begin{cases} 1 & n = 0 \\ 4 & n = 1 \\ 8 & n = 2 \\ 6 & n \geq 3. \end{cases}$$

A similar computation as before shows that the growth series with respect to this generating set is also rational:

$$\mathcal{S}_{\{2,3\}}(z) = 1 + 4z + 8z^2 + 6z^3 + 6z^4 + \cdots = \frac{1 + 3z + 4z^2 - 2z^3}{1 - z}.$$

We see that the growth series with respect to both generating sets are rational and they have the same denominator. The following exercise asks you to add a bit of evidence in support of the claim that, in the case of an infinite cyclic group, the growth series with respect to any generating set is rational with denominator $1 - z$.

**Exercise 9.12.** Let $S$ be a pair of relatively prime integers, so that $S$ forms a finite generatoring set for $\mathbb{Z}$. Show that the associated growth series is rational, with denominator $1 - z$.

**Example 9.13.** Let $W_{333}$ be the group generated by three reflections in the sides of an equilateral triangle, discussed in Chapter 2. The Cayley graph embeds in the plane as the edges of the triangulation by regular hexagons. Thus it suffices to compute how many vertices are a distance $n$ from a fixed base vertex in this graph. Or one can count the number of triangles separated from a base triangle by at most $n$ edges in the tiling by equilateral triangles. In either approach one quickly comes to the conjecture that

$$\sigma(n) = \begin{cases} 1 & n = 0 \\ 3 \cdot n & n > 0. \end{cases}$$

This formula is correct, but it takes some care to actually prove it. One can also show that the corresponding growth series is the formal power series associated to the rational function

$$\mathcal{S}(z) = \frac{1 + z + z^2}{(1 - z)^2}.$$

(Exercises 5 and 6 ask you to verify these claims.)

Below are two theorems that compute growth series of products of groups, assuming you know the growth series of the individual factors.

**Theorem 9.14.** *Let $G$ and $H$ be groups with finite generating sets $S_G$ and $S_H$, and corresponding growth series $\mathcal{S}_G(z)$ and $\mathcal{S}_H(z)$. Then the set*

$$S_{G \oplus H} = \{(s, e_H) \mid s \in S_G\} \cup \{(e_G, h) \mid h \in S_H\}$$

*is a generating set for $G \oplus H$ and the corresponding growth series for $G \oplus H$ is given by*

$$\mathcal{S}_{G \oplus H}(z) = \mathcal{S}_G(z) \cdot \mathcal{S}_H(z).$$

*Proof.* The claim about generating sets is clear.

The length of $(g, h) \in G \oplus H$ is the sum of the lengths of $g$ and $h$. So

$$\sigma_{G \oplus H}(n) = \sum_{i=0}^{n} \sigma_G(i)\sigma_H(n - i),$$

from which the statement follows.                                    □

An induction argument establishes the following result:

**Corollary 9.15.** *The growth series for $\mathbb{Z}^n$ with respect to a standard generating set is*

$$\mathcal{S}_{\mathbb{Z}^n}(z) = \left(\frac{1 + z}{1 - z}\right)^n.$$

**Theorem 9.16.** *Let $G$ and $H$ be groups with finite generating sets $S_G$ and $S_H$, and corresponding growth series $\mathcal{S}_G(z)$ and $\mathcal{S}_H(z)$. Take as generating set for $G * H$ the image of $S_G$ and $S_H$ in $G * H$. Then*

$$\frac{1}{\mathcal{S}_{G*H}(z)} = \frac{1}{\mathcal{S}_G(z)} + \frac{1}{\mathcal{S}_H(z)} - 1.$$

*Proof.* Every element in $G * H$ can be expressed as

$$g_0 \cdot \underbrace{h_1 g_1 h_2 g_2 \cdots h_{n-1} g_{n-1} h_n g_n}_{\text{none equal to the identity}} \cdot h_{n+1}$$

where $g_0$ and/or $h_{n+1}$ may be equal to the identity. It follows that

$$\begin{aligned}
\mathcal{S}_{G*H}(z) &= \sum_{n \geq 0} \mathcal{S}_G(z) \left[ (\mathcal{S}_H(z) - 1)(\mathcal{S}_G(z) - 1) \right]^n \mathcal{S}_H(z) \\
&= \frac{\mathcal{S}_G(z) \cdot \mathcal{S}_H(z)}{1 - (\mathcal{S}_H(z) - 1)(\mathcal{S}_G(z) - 1)}
\end{aligned}$$

where the last equality is established using the standard formula for geometric series. A bit of algebra converts this formula into the one stated in the theorem.                                                    □

**Example 9.17.** Pick cyclic generators for $\mathbb{Z}_3$ and $\mathbb{Z}_4$ and the corresponding growth "series" for these finite groups are $1+2z$ and $1+2z+z^2$. Applying Theorem 9.16 we see that the growth series for $\mathbb{Z}_3 * \mathbb{Z}_4$ is

$$\begin{aligned}
\mathcal{S}_{\mathbb{Z}_3 * \mathbb{Z}_4}(z) &= \frac{(1+2z)(1+2z+z^2)}{1 - (2z)(2z+z^2)} = \frac{1 + 4z + 5z^2 + 2z^3}{1 - 4z^2 - 2z^3} \\
&= 1 + 4z + 9z^2 + 20z^3 + 44z^4 + 98z^5 + 216z^6 + \cdots.
\end{aligned}$$

**Remark 9.18.** Not all growth series of finitely generated groups are rational. For example, the growth series of $\mathbb{Z}_2 \wr \mathbb{F}_2$, a "lamplighter" group based on lighting lamps on the Cayley graph of $\mathbb{F}_2$ instead of $\mathbb{Z}$, is not rational. (See [Pa92] for the details on this and related examples.) The finitely generated, infinite torsion group $\mathcal{U}$, which we constructed in Chapter 6, is another example. Perhaps a bit more worrisome is the fact that the growth series can be rational with respect to one generating set but not rational with respect to another [St96].

## 9.3 Growth and Regular Languages

There is a connection between the idea of growth series and groups which admit regular normal forms.

**Theorem 9.19.** *Let $G$ be a group with $S$ a finite set of generators. If $(G, S)$ admits a regular normal form that is composed of geodesic words, then the growth series for $G$ with respect to $S$ is rational.*

The proof of this theorem uses the "transfer-matrix method," a tool of broad utility in combinatorial arguments. In fact, this result really has little to do with groups; it is a theorem about the growth series of regular languages.

Let $\mathcal{G}$ be a finite directed graph. The *adjacency matrix* $A$ has as its $(i, j)$-entry the number of edges connecting vertex $i$ to vertex $j$ in $\mathcal{G}$. Let $A(n)$ be the matrix whose $(i, j)$-entry is the number of directed edge paths of length exactly $n$, which connect vertex $i$ to vertex $j$.

**Lemma 9.20.** *The matrix $A(n)$ equals $A^n$.*

*Proof.* This is what matrix multiplication is all about! (If that doesn't make you happy, prove it by induction.) □

**Lemma 9.21.** *Let $\mathcal{S}_{ij}(z) = \sum_{n \geq 0} A_{ij}(n) z^n$. Then $\mathcal{S}_{ij}(z)$ is the series corresponding to the rational function*

$$\mathcal{S}_{ij}(z) = \frac{(-1)^{i+j} \det (I - zA : j, i)}{\det (I - zA)}$$

*where $(I - zA : j, i)$ denotes the matrix formed by subtracting $zA$ from the identity matrix, and then removing the $j$th row and $i$th column.*

*Proof.* The series $\mathcal{S}_{ij}(z)$ is the $(i, j)$-entry of

$$\sum_{n \geq 0} A^n z^n = \sum_{n \geq 0} (zA)^n = (I - zA)^{-1} .$$

Thus the formula stated is simply the standard formula for the inverse of a matrix. □

*Proof of Theorem 9.19.* Let $\mathcal{M}$ be a finite automaton, with no $\epsilon$-edges, whose language is the regular language of normal forms for $G$. Let $A$ be the associated adjacency matrix. Then, by the previous two lemmas, we get that the rational function associated to the growth series is given by

$$S \cdot (I - zA)^{-1} \cdot A$$

where: $S$ is a $1 \times |V(\mathcal{M})|$ matrix with 1s in each spot associated to a start state, and zeros elsewhere; and $A$ is a $|V(\mathcal{M})| \times 1$ matrix with 1s corresponding to accept states, and zeros elsewhere. □

**Example 9.22.** The collection of freely reduced words is a normal form for $\mathbb{F}_2$. This language is also regular and geodesic. An automaton whose language consists of the freely reduced words in $\{x, y, x^{-1}, y^{-1}\}^*$ is shown in Figure 7.5. This machine has five states, consisting of a start state, and then four states corresponding to the last letter read. The associated adjacency matrix, with the states organized as indicated, is:

$$
\begin{array}{l}
\text{Start state} \\
\text{Last read } x \\
\text{Last read } y \\
\text{Last read } x^{-1} \\
\text{Last read } y^{-1}
\end{array}
\left(
\begin{array}{ccccc}
0 & 1 & 1 & 1 & 1 \\
0 & 1 & 1 & 0 & 1 \\
0 & 1 & 1 & 1 & 0 \\
0 & 0 & 1 & 1 & 1 \\
0 & 1 & 0 & 1 & 1
\end{array}
\right).
$$

The formula given in the proof of Theorem 9.19 is then

$$
\begin{pmatrix} 1 & 0 & 0 & 0 & 0 \end{pmatrix}
\begin{pmatrix}
1 & -z & -z & -z & -z \\
0 & 1-z & -z & 0 & -z \\
0 & -z & 1-z & -z & 0 \\
0 & 0 & -z & 1-z & -z \\
0 & -z & 0 & -z & 1-z
\end{pmatrix}^{-1}
\begin{pmatrix} 1 \\ 1 \\ 1 \\ 1 \\ 1 \end{pmatrix}.
$$

This may look daunting, but a program such as *Mathematica* can do the computation almost instantly, giving the answer:

$$
\frac{1+z}{1-3z} = 1 + 4z + 12z^2 + 36z^3 + 108z^4 + \cdots.
$$

Since $\sigma(n) = 4 \cdot 3^{n-1}$ for $\mathbb{F}_2$ with respect to a basis, the reader should be able to independently verify that this formula is indeed correct.

The method described above involves one step that is computationally intensive, namely, the inversion of a matrix with polynomial entries. While *Mathematica* can handle the computation in the case considered in Example 9.22, in more complicated situations such programs have great difficulty computing inverses. One can, however, avoid this step. The inverse of $I - zA$ can be expressed as

$$
(I - zA)^{-1} = \frac{1}{\det [I - zA]} M
$$

where $M$ is a matrix with polynomial entries. It follows that

$$
S \cdot (I - zA)^{-1} \cdot A = \frac{p(z)}{\det [I - zA]}
$$

where $p(z)$ is some polynomial. Further, by appealing to the formula in Lemma 9.21, we see that the degree of $p(z)$ must be no greater than $(n-1)$ if $A$ is an $n$-by-$n$ matrix.

Revisiting Example 9.22, we see that the determinant of $I - zA$ is $(z - 1)^2(1 + z)(1 - 3z)$. Thus we know the growth series can be expressed as

$$\frac{a + bz + cz^2 + dz^3 + ez^4}{(z-1)^2(1+z)(1-3z)} = a + (4a + b)z$$
$$+ (14a + 4b + c)z^2 + (44a + 14b + 4c + d)z^3$$
$$+ (135a + 44b + 14c + 4d + e)z^4 + \cdots,$$

where we have used *Mathematica* to produce the expansion. Since we know the constant term in the growth series is 1, we have $a = 1$. The coefficient of $z$ is just the number of generators (and their inverses), so $4a + b = 4$, hence $b = 0$. Continuing in this manner we can quickly determine the numerator, and hence the rational function.

**Remark 9.23.** It is known that the growth series for BS$(1,2)$, with respect to the standard generators $a$ and $b$, is rational, but where the rational function has degree 14. That the growth series is rational was established by Brazil in [Bz94] and independently by Collins, Edjvet and Gill [CEG94].[2] However, it has also been proved that BS$(1,2)$ does not admit a regular normal form consisting of geodesic words [Gv96]. In particular, it is possible to have a rational growth series without it being computed via the method described in this section.

All three of the papers cited above are accessible to anyone who has read this far in this book, and are recommended for further study.

Finally we should point out that connections between numerical sequences, formal power series and rational functions has been well-studied by combinatorialists. A good introduction to the topic of "generating functions" is [Wi06].

## 9.4 Cannon Pairs

Fix a group $G$ and a finite generating set $S$. Let $\gamma(n)$ be the number of geodesic words of length $n$, and let $\gamma(g)$ be the number of geodesic

---

[2] The formula for the rational function mentioned above was not computed by Brazil as it would be "complicated and tedious"; Collins, Edjvet and Gill arrived at the result independently and worked through the details to arrive at a final formula.

expressions for the element $g \in G$. The *geodesic growth series* is

$$\Gamma(z) = \sum_{n \geq 0} \gamma(n) z^n = \sum_{g \in G} \gamma(g) z^{|g|}.$$

**Example 9.24.** Consider $\mathbb{Z} \oplus \mathbb{Z}$ with generators $\{(1,0),(0,1)\}$. If $n$ and $m$ are two positive integers, then the number of geodesics expressing the element $(m,n)$ is $\binom{n+m}{n}$. The number of geodesics of length $n$ that occur in the first quadrant is then

$$\sum_{i=0}^{n} \binom{n}{i} = \binom{n}{0} + \binom{n}{1} + \cdots + \binom{n}{n-1} + \binom{n}{n} = 2^n.$$

Adding the four quadrants, and subtracting 4 for the elements on the axes that have been counted twice shows $\gamma(n) = 4(2^n - 1)$ for $n \geq 1$. Thus

$$\begin{aligned}
\Gamma(z) &= 1 + 4z + 12z^2 + 28z^3 + 60z^4 + \cdots \\
&= 1 + [8z + 16z^2 + 32z^3 + \cdots] - [4z + 4z^2 + 4z^3 + \cdots] \\
&= 1 + 4 \left[ \frac{2z}{1-2z} \right] - 4 \left[ \frac{z}{1-z} \right] = \frac{1 + z + 2z^2}{(1-z)(1-2z)}.
\end{aligned}$$

This example indicates that the geodesic growth series of a group, much like the ordinary growth series of a group, may be rational. An argument similar to that given in Section 9.3, using the transfer-matrix method, establishes:

**Theorem 9.25.** *Let $(G, S)$ be a group paired with a finite generating set. If the language of all geodesic words is a regular language, then the geodesic growth series is rational.*

In 1984, James Cannon published a foundational paper for the study of infinite groups from a geometric perspective [Ca84]. In particular, he imported the idea of cone types, which we have encountered in our discussion of formal languages, to the study of geodesics in the Cayley graphs of infinite groups.

**Definition 9.26.** Fix a group $G$ and a finite generating set $S$, and let $\Gamma = \Gamma_{G,S}$ be the associated Cayley graph. Then, for a given $g \in G$,

$\mathrm{Cone}(g) = $ The subgraph of $\Gamma$ induced by

$$\{v_h \mid v_g \text{ is on a geodesic from } v_e \text{ to } v_h\}.$$

The elements $g$ and $h \in G$ are said to have the *same cone type* if there is an edge-label preserving isomorphism between the directed graphs

Cone($g$) and Cone($h$). This is an equivalence relation, and if there are only finitely many equivalence classes then $G$ is said to have *finitely many cone types*.

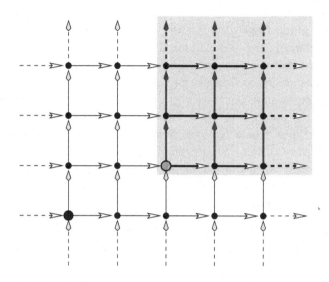

Fig. 9.2. The lower-left vertex is the vertex associated to $(0,0) \in \mathbb{Z} \oplus \mathbb{Z}$ while the highlighted vertex is associated to $(2,1) \in \mathbb{Z} \oplus \mathbb{Z}$. The cone of this element is the highlighted subgraph, which looks much like a quadrant in the Euclidean plane.

**Example 9.27.** Consider $\mathbb{Z} \oplus \mathbb{Z}$ with the generators $\{(1,0),(0,1)\}$. The cone of the element $(2,1)$ consists of the subgraph of the Cayley graph that is induced by $\{(m,n) \mid m \geq 2 \text{ and } n \geq 1\}$. This is illustrated in Figure 9.2.

**Theorem 9.28.** *Let $G$ be a group with finite generating set $S$. Then $(G,S)$ has finitely many cone types if and only if the language of all geodesic words is regular.*

*Proof.* In Section 7.2 we defined "cone type" in the context of languages:

Let $\mathcal{L}$ be a language in $\mathcal{S}^*$ and let $\omega \in \mathcal{S}^*$. The *cone type* of $\omega$ is the set of all words $\omega' \in \mathcal{S}^*$ such that $\omega \cdot \omega' \in \mathcal{L}$. Denote the cone type of $\omega$ by Cone($\omega$). The collection

$$\text{Cone}(\mathcal{L}) = \{\text{Cone}(\omega) \mid \omega \in \mathcal{S}^*\} \,.$$

is the *cone type of the language $\mathcal{L}$*, denoted Cone($\mathcal{L}$).

To avoid confusing the different uses of this phrase, we refer to the cone types of $\mathcal{L}$ as $\mathcal{L}$-*Cone types*.

Let $\mathcal{L}$ be the language of all geodesic words in $\{S \cup S^{-1}\}^*$. If $\mathcal{L}$ is regular, then the Myhill–Nerode Theorem (7.18) states that there are finitely many $\mathcal{L}$-cone types. Given an element $g \in G$, the subgraph of the Cayley graph Cone($g$) is just the union of paths, based at $g$, corresponding to words in $\mathcal{L}$-Cone($\omega$) for all geodesic words $\omega$ with $\pi(\omega) = g$. As there are finitely many sets of $\mathcal{L}$-Cones, there can only be finitely many graphs arising as Cone($g$) for $g \in G$.

Conversely, assume the Cayley graph $\Gamma_{G,S}$ has finitely many Cones. Let $\mathcal{M}$ be the automaton whose vertex set corresponds to the finitely many cone types, where the cone type of the identity is the only start state, and every vertex is an accept state. For each generator $s \in S$ add a directed edge labelled $s$ from Cone($g$) to Cone($gs$) when the vertex associated to $gs$ is in Cone($g$). The language accepted by this machine is $\mathcal{L}$, hence $\mathcal{L}$ is regular. $\qquad\square$

**Definition 9.29.** If the pair $(G, S)$ has finitely many cone types, or equivalently if the language of all geodesics is regular, then $(G, S)$ is a *Cannon pair*.

In the remainder of this section we outline an example due to Jim Cannon. Let $x$ and $y$ be the standard generators of $\mathbb{Z} \oplus \mathbb{Z}$ and let $\alpha \in$ Aut($\mathbb{Z} \oplus \mathbb{Z}$) be the automorphism induced by $\alpha(x) = y$ and $\alpha(y) = x$. The automorphism $\alpha$ has order 2, hence we can use it to construct a semi-direct product $G = (\mathbb{Z} \oplus \mathbb{Z}) \rtimes \mathbb{Z}_2$.

For the remainder of this example we will slightly abuse notation, and let $x = [x, e], y = [y, e]$ and $\alpha = [e, \alpha]$. The Cayley graph of $G$ with respect to the generating set $\{x, y, \alpha\}$ consists of parallel grids. The subgraph induced by $\{x, y\}$ is the standard Cayley graph of $\mathbb{Z} \oplus \mathbb{Z}$. There are (unoriented) edges labelled $\alpha$ joining $[x^m y^n, e]$ to $[x^m y^n, \alpha]$ for all $m, n \in \mathbb{Z}$. So it only remains to understand how right multiplication by $x$ and $y$ influences $[x^m y^n, \alpha]$. This is relatively straightforward, since

$$[x^m y^n, \alpha] \cdot [x, e] = [x^m y^n \alpha(x), \alpha \cdot e] = [x^m y^{n+1}, \alpha],$$

and similarly

$$[x^m y^n, \alpha] \cdot [y, e] = [x^m y^n \alpha(y), \alpha \cdot e] = [x^{m+1} y^n, \alpha].$$

Thus the subgraph containing $[e, \alpha]$, whose edges are all labelled $x$ and $y$, is a copy of the $\mathbb{Z} \oplus \mathbb{Z}$ Cayley graph with the roles of $x$ and $y$ reversed.

Exercise 13 asks you to verify that this Cayley graph has finitely many cone types.

We now switch our generating set. Define $d = xy$ and $z = x^2$ and consider the set of elements $\widehat{S} = \{x, d, z, \alpha\} \in G$. Since $x$ and $d$ are in $\widehat{S}$, $y = x^{-1}d \in \{\widehat{S} \cup \widehat{S}^{-1}\}^*$. Thus $\widehat{S}$ is a generating set for $G$. Exercise 14 asks you to prove that, in this generating set, the word $\alpha z^n \alpha z^m$ is geodesic if and only if $m < n$. Thus the language of geodesics is not regular with respect to this generating set by an application of the Pumping Lemma (Theorem 7.14).

Because of examples like this, it is important to remember that having the language of all geodesics be regular is a property of a *pair* consisting of a group and a fixed generating set. It is not a property of the group independent of the choice of generators.

## 9.5 Cannon's Almost Convexity

Jim Cannon introduced the following property in 1987.

**Definition 9.30** (Almost convex). A Cayley graph is *almost convex* if there is a uniform constant $K \geq 0$ such that if $v, w \in \mathcal{S}_\Gamma(n)$, and $d_\Gamma(v, w) \leq 2$, then $d_{\mathcal{B}(n)}(v, w) \leq K$. In other words, a Cayley graph $\Gamma$ is almost convex if whenever two vertices on the sphere of radius $n$ can be joined by a path of length $\leq 2$ in $\Gamma$ then they can be joined by a path of length $\leq K$ that stays inside the ball of radius $n$. (As this is a property of the Cayley graph, the "balls" under consideration are balls in the Cayley graph.)

A group is *almost convex* if it has a Cayley graph that is almost convex.

**Example 9.31.** Free abelian groups are almost convex as are free groups and free products of finite groups. Each of these claims is true regardless of the set of generators that is chosen, but these claims are easiest to verify using "standard" generators. Showing this is true for any set of generators is actually somewhat difficult. There are, in fact, groups whose Cayley graphs are almost convex with respect to one set of generators, but not with respect to another.

Note: In the ball of radius $n$ you can always join two vertices by a path of length $\leq 2n$. You simply move from one vertex to the identity, and then back out to the other vertex. The power of the almost convexity condition is that there is a single, uniform constant that works for all $n$.

It should be noted that if there is such a constant for vertices a distance at most 2 apart, then there is also a uniform constant when the vertices are a distance 3 apart, or 4 apart, or ...

**Proposition 9.32.** *If $\Gamma$ is almost convex, then given any $m \geq 2$ there is a fixed constant $K(m)$ such that if $v, w \in S_\Gamma(n)$, and $d_\Gamma(v, w) \leq m$, then $d_{\mathcal{B}(n)}(v, w) \leq K(m)$.*

*Proof.* Let $K = K(2)$. We argue by induction, the base case being our hypothesis that $\Gamma$ is almost convex. Assume then that $v, w \in S_\Gamma(n)$, and $d_\Gamma(v, w) = m$, where $m \geq 3$. Let $M$ be the maximum distance from the vertex associated to the identity to a vertex on the chosen edge path joining $v$ to $w$. (Thus the chosen path is contained in $\mathcal{B}_\Gamma(M + 1)$.) Note that $M$ must be less than $n + \lfloor \frac{m+1}{2} \rfloor$.

Let $z$ be a vertex on this path where the distance from $z$ to the vertex associated to the identity is $M$. The neighboring vertices on this path, call them $z_1$ and $z_2$, are then either closer to the identity, or at least one of them is the same distance from the identity as $z$. If both $z_1$ and $z_2$ are in $S_\Gamma(M-1)$ then they may be joined by a path inside of $\mathcal{B}_\Gamma(M-1)$ whose length is at most $K$. This removes a vertex in $S_\Gamma(M)$. In the other case, $z$ and $z_1$ (or $z_2$) can be joined by a path inside of $\mathcal{B}_\Gamma(M)$. This removes an edge that was outside of $\mathcal{B}_\Gamma(M)$. In a worse-than-is-possible scenario, this process would have to be applied $m$ times, rerouting the original path from its original vertices or edges, resulting in a path joining $v$ to $w$ of length $mK$. This new path is contained in $\mathcal{B}_\Gamma(M)$.

Repeating this process results in an edge path of length less than $(mK)K = mK^2$, which is contained in $\mathcal{B}_\Gamma(M - 1)$. Continuing this process we eventually convert the original path to one which is contained in $\mathcal{B}_\Gamma(n)$. Since $(M + 1) - n \leq 1 + \lfloor \frac{m+1}{2} \rfloor$, if we set $l = 1 + \lfloor \frac{m+1}{2} \rfloor$ then this process stops in at most $l$ steps. The length of the path at the end can be no more than $mK^l$, hence we may take $K(m) = mK^l$.          $\square$

One reason to be interested in almost convex groups is:

**Theorem 9.33.** *If $G$ is almost convex, then it has a solvable word problem.*

*Proof.* We will prove that if $G$ is almost convex then there is an algorithm that allows you to construct its (almost convex) Cayley graph $\Gamma$. The argument is, however, non-constructive. In other words, we prove that the algorithm exists, but we do not actually provide it! The argument begins by mentioning that $\mathcal{B}_\Gamma(K+2)$ exists, where $K$ is the almost convexity constant. Let $C$ be any circuit in $\Gamma$ of length $\leq K+2$, and let $v_g$ be a vertex in this circuit. Then by multiplying by $g^{-1}$ we may translate this circuit into $\mathcal{B}_\Gamma(K+2)$. Thus all circuits of length $\leq K+2$ in $\Gamma$ can be realized as translates of circuits in $\mathcal{B}_\Gamma(K+2)$.

Assuming you have built $\mathcal{B}_\Gamma(n)$, you can make a list of "missing edges," that is, edges which are attached to vertices in $S_\Gamma(n)$ but which are not contained in $\mathcal{B}_\Gamma(n)$. (See the proof of Theorem 5.10.) We need to determine how to attach these missing edges in order to form $\mathcal{B}_\Gamma(n+1)$. If a missing edge joins two vertices in $S_\Gamma(n)$ then those vertices must be joined by a path of length $\leq K$ in $\mathcal{B}_\Gamma(n)$. Similarly, if two missing edges meet at a vertex in $S_\Gamma(n+1)$ then their associated vertices in $S_\Gamma(n)$ are joined by a path of length $\leq K$ in $\mathcal{B}_\Gamma(n)$. In either event, there is then a circuit formed by the missing edge, or pair of edges, and an edge path of length $\leq K$. If there is such a circuit, a copy of it can be found in $\mathcal{B}_\Gamma(K+2)$. If no such circuit appears in $\mathcal{B}_\Gamma(K+2)$ then the missing edges go to distinct vertices in $S_\Gamma(n+1)$.

Since there is an algorithm which constructs the Cayley graph $\Gamma$, then by Theorem 5.10 there is an algorithm that solves the word problem. $\square$

As we mention below, not all groups are almost convex. There are a number of weaker versions of almost convexity that have attracted interest. The weakest (useful) variation is *minimal almost convexity*. A Cayley graph $\Gamma$ is *minimally almost convex* if, given $v, w \in S_\Gamma(n)$, where $d_\Gamma(v, w) \leq 2$, then $d_{\mathcal{B}(n)}(v, w) \leq 2n - 1$. (Remember, $d_{\mathcal{B}(n)}(v, w)$ is always less than or equal to $2n$.) A finitely generated group is then minimally almost convex if it has a minimally almost convex Cayley graph.

**Example 9.34.** It was shown in 1998 that $BS(1, 2)$ is not almost convex with respect to any finite generating set [MiS98]. (Exercise 16 gives a hint as to how to prove that the Cayley graph of $BS(1, 2)$, with respect to the generators given in Chapter 4, is not almost convex.) However, in 2005 it was shown that $BS(1, 2)$ is minimally almost convex [EH05].

There are groups that are not minimally almost convex. For example, Cleary and Taback proved that the lamplighter group $L_2$ is not minimally almost convex [CT05]. Their approach is to consider the elements $[\{-n, n\}, 1]$ and $[\{-n, n\}, -1]$. By Proposition 9.7, the length of both of these elements is $4n + 1$. They are both adjacent to $[\{-n, n\}, 0]$, hence the distance between them is 2. However, the intermediate element, $[\{-n, n\}, 0]$, has length $4n + 2$, hence it is not in the ball of radius $4n + 1$.

In order to get from $[\{-n, n\}, -1]$ to $[\{-n, n\}, 1]$ the location of the lamplighter must at some point be 0. If both lamps at $-n$ and $n$ are lit, then having the lamplighter pass though 0 requires that you move outside of the ball of radius $4n + 1$. Thus to get from $[\{-n, n\}, -1]$ to $[\{-n, n\}, 1]$, staying inside the ball of radius $4n + 1$, one must first dowse the lamp at $-n$. This requires a minimum of $n$ steps. In the end, the lamp at $-n$ will need to be lit, but it cannot be lit until the lamp at $n$ is dowsed. Thus a minimum of $2n + 1$ more steps need to be taken in order to turn off the lamp at $n$. Because the lamplighter ends up to the right of the origin, we need to light the lamp at $-n$ before lighting the lamp at $n$. So we again need a minimum of $2n + 1$ steps to return to $-n$ and light the lamp. Finally, in order to return to the right side and light the lamp at $n$ an additional $2n + 1$ steps are needed, and moving from $n$ to 1 requires $n - 1$ steps. Thus, no matter how clever you are in choosing your path, any path from $[\{-n, n\}, -1]$ to $[\{-n, n\}, 1]$ that is contained in the ball of radius $4n + 1$ must have length at least $8n + 2 = 2(4n + 1)$. Thus we see the following result:

**Proposition 9.35.** *The Cayley graph of the lamplighter group with respect to the generators $t = [\emptyset, 1]$ and $a = [\{0\}, 0]$ is not minimally almost convex.*

## Exercises

(1)  Let $x$ generate a cyclic group of order 5. Show that the length of $x^2$ and the length of $x^3$ (with respect to $\{x, x^{-1}\}$ are both equal to 2.

(2)  Characterize the set of elements in $L_2$ for whom the right-first normal form is geodesic.

(3)  Let $G$ be a group with generating set $\mathcal{S} = S \cup S^{-1}$. The corresponding notion of length induces a function $L : G \to \mathbb{N}$ given by $L(g) = |g|$. These length functions are surprisingly easy to axiomatize. Let $\ell : G \to \mathbb{N}$ be a function. Prove that $\ell$ is the same as $L$ if and only if

   a.  $\ell(e) = 0$;

   b.  $|\ell(gs) - \ell(g)| \leq 1$ for all $g \in G$ and $s \in \mathcal{S}$; and

   c.  if $g \in G \setminus \{e\}$, then there is at least one $s \in \mathcal{S}$ such that $\ell(gs) < \ell(g)$.

(4)  Find a formula for the size of $\mathcal{S}(n)$ and $\mathcal{B}(n)$ for $\mathbb{Z} \oplus \mathbb{Z}$ with respect to the standard generating set $(1, 0)$ and $(0, 1)$. Then find a formula for the size of $\mathcal{S}(n)$ and $\mathcal{B}(n)$ with respect to the larger generating set $\{(1, 0), (0, 1), (1, 1)\}$.

(5)  Let $W_{333}$ be the group generated by three reflections in the sides of an equilateral triangle, discussed in Chapter 2. Prove that $\sigma(n) = |\mathcal{S}(n)|$ is given by

$$\sigma(n) = \begin{cases} 1 & n = 0 \\ 3 \cdot n & n > 0. \end{cases}$$

(6)  Prove that the growth series for $W_{333}$ is the formal power series associated to the rational function

$$\mathcal{S}(z) = \frac{1 + z + z^2}{(1 - z)^2}.$$

(7)  Theorem 9.16 gives a formula for the growth series of a free product of two groups. Find a similar formula for $G \approx A * B * C$.

(8)  The growth series for finitely generated free groups, with respect to a basis, are always rational. In Section 9.2 it is shown that the growth series for $\mathbb{F}_1$ is the series associated to $(1+z)/(1-z)$; in Section 9.3 we saw that the growth series for $\mathbb{F}_2$ is the series associated to $(1+z)/(1-3z)$. What is the rational function corresponding to the growth series for $\mathbb{F}_n$?

(9)  Let $\{a, b\}$ be the generating set for $BS(1, 2)$ described in Chapter 4. Show that the length of $b^n$ $(n \geq 1)$ is obtained by writing $n$ in binary: If $n = b_k b_{k-1} \cdots b_2 b_1 b_0$ (in binary), with $b_k \neq 0$, then $|b^n|$ is $2k + \sum_{i=0}^{k} b_i$. For example, $14 = 8 + 4 + 2$ so, written in binary, 14 is 1110. The claim is that $|b^{14}|$ is $2 \cdot 3 + 1 + 1 + 1 = 9$.

(10)  If $\sigma(n) = |\mathcal{S}(e, n)|$, then the associated series $\mathcal{S}(z) = \sum_{n \geq 0} \sigma(n) z^n$ is sometimes referred to as the *spherical* growth series. One can also consider the function $\beta(n) = |\mathcal{B}(e, n)|$ which gives the number of vertices inside the ball of radius $n$, and the associated *ball* growth series $\mathcal{B}(z) = \sum_{n \geq 0} \beta(n) z^n$. Show that the spherical growth series is rational if and only if the ball growth series is rational.

(11)  Given any regular language $\mathcal{L}$ over an alphabet $A$, one can form the associated growth series $\sum_{w \in \mathcal{L}} z^{|w|}$. The proof of Theorem 9.19 establishes that, if $\mathcal{L}$ is regular, this growth series is rational. Show that the converse is false using the language $\mathcal{L}$ of all palindromes over the alphabet $A = \{a, b\}$.

(12) Let $G = \mathbb{Z} \oplus \mathbb{Z}$ and let $S = \{(1,0),(0,1)\}$ be the standard set of generators. Show that $(G, S)$ has exactly nine cone types.

(13) Let $G = (\mathbb{Z} \oplus \mathbb{Z}) \rtimes \mathbb{Z}_2$ and let $S$ be the set of generators

$$S = \{[(1,0), e], [(0,1), e], [(0,0), \alpha]\}$$

(as discussed in Section 9.4). Prove that $(G, S)$ has finitely many cone types, that is, that $(G, S)$ is a Cannon pair.

(14) Let $G$ be as in Exercise 13 and define $d = xy$ and $z = x^2$. Show that in this generating set the word $\alpha z^n \alpha z^m$ is geodesic if and only if $m < n$.

(15) Let $G$ and $H$ be almost convex groups. Show that $G \oplus H$ and $G * H$ are also almost convex groups.

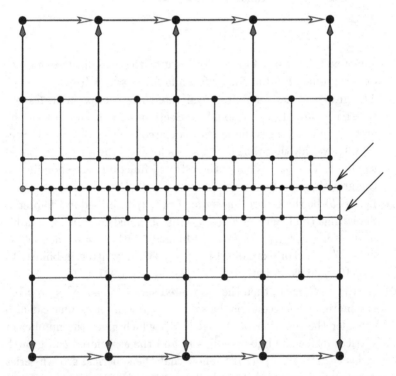

Fig. 9.3. Two vertices in the sphere of radius 10 in the Cayley graph of BS$(1, 2)$ are highlighted. They are a distance 2 apart in the Cayley graph but are much further apart inside the ball of radius 10.

(16) This exercise indicates how to prove that the Cayley graph of BS$(1, 2)$, with respect to the generators $\{a, b\}$, is not almost convex. The reader may want to refer to the view of the Cayley graph of BS$(1, 2)$ shown in Figure 9.3 throughout this problem.

a. Show that $ab^4a^{-1}$ and $bab^4$ are both in the sphere of radius 6, and that the distance between them in the Cayley graph is 2.

b. What is the distance between $ab^4a^{-1}$ and $bab^4$ when restricted to paths in the ball of radius 6?

c. Show that $a^2b^4a^{-2}$ and $ba^2b^4a^{-1}$ are both in the sphere of radius 8, and that the distance between them in the Cayley graph is 2.

d. What is the distance between $a^2b^4a^{-2}$ and $ba^2b^4a^{-1}$ when restricted to paths in the ball of radius 8?

e. Show that $a^3b^4a^{-3}$ and $ba^3b^4a^{-2}$ are both in the sphere of radius 10, and that the distance between them in the Cayley graph is 2. (These are the vertices highlighted in Figure 9.3.)

f. What is the distance between $a^3b^4a^{-3}$ and $ba^3b^4a^{-2}$ when restricted to paths in the ball of radius 10?

g. Attempt to prove that the Cayley graph of BS(1, 2), with respect to $\{a, b\}$, is not almost convex. If you get stuck, try reading [MiS98] or [EH05].

(17) Given a group $G$ and a finite generating set $S$, a *dead end* is an element $g$ such that $|gs| \le |g|$ for all $s \in \{S \cup S^{-1}\}$. Finite groups always have dead end elements, but it is somewhat surprising that infinite groups can have dead end elements as well.

Dead end elements are also sometimes referred to as *pockets*. (If the Cayley graph $\Gamma$ is thought of as a net, dangling down from the vertex associated to the identity, then a dead end element sits at the bottom of a "pocket.") The *depth* of a dead end element $g$ is

$$\text{Depth}(g) = \text{Min}\{d(g, h) \mid |h| > |g|\}.$$

In other words, $g$ has depth $d$ if one needs to take $d$ steps in the Cayley graph in order to get from $g$ to an element further from the identity than $g$. The Cayley graph $\Gamma_{G,S}$ is said to have *deep pockets* if there is $n \in \mathbb{N}$ such that $n \ge \text{Depth}(g)$ for all dead ends $g \in G$. Dead ends in $L_2$ and other wreath products are studied in [CT05].

Let $d = [\{-1, 0, 1\}, 0] \in L_2$. Show that $|d| = 7$, but $|d \cdot t| = |d \cdot t^{-1}| = |d \cdot a| = 6$, and therefore $d \in L_2$ is a dead end. Also show that the depth of $d$ is 2.

(18) Let $d_n \in L_2$ be the element whose picture has lit lamps at each integer from $-n$ to $n$, with the lamplighter positioned at the origin. Show that $d_n$ is a dead end element of depth $n + 1$, and conclude that the Cayley graph of $L_2$ with respect to the generators $\{a, t\}$ has deep pockets.

# 10

# Thompson's Group

Thompson's group comes up in many different situations, it has interesting properties, and almost every question you can ask about it is challenging.

–Kenneth S. Brown

In this chapter we introduce another interesting infinite group described in terms of functions from $\mathbb{R}$ to $\mathbb{R}$, although in this case we only use the closed interval $[0, 1]$. We begin by discussing dyadic[1] divisions of $[0, 1]$. A *dyadic division* of $[0, 1]$ is constructed by first dividing $[0, 1]$ into $[0, 1/2]$ and $[1/2, 1]$, and then proceeding to "pick middles" of the resulting pieces, a finite number of times. For example,

$$[0, 1] = [0, 1/2] \cup [1/2, 5/8] \cup [5/8, 3/4] \cup [3/4, 1]$$

is a dyadic division. We refer to the points of subdivision as the *chosen middles*, so in the example above the chosen middles are: $1/2, 3/4$ and $5/8$ (listed in the order they are chosen). We refer to any interval of the form $\left[\dfrac{m}{2^n}, \dfrac{m+1}{2^n}\right]$ ($0 \leq m \leq 2^n - 1$) as a *standard dyadic interval*. Exercise 1 asks you to show that any time you divide $[0, 1]$ into standard dyadic intervals, you necessarily have a dyadic division of $[0, 1]$.

Dyadic divisions of $[0, 1]$ can be encoded using finite, rooted binary trees. Recall from Chapter 6 that a rooted binary tree is a rooted tree where every non-leaf vertex has two descendants. Let $\mathcal{T}$ be the rooted tree whose vertex set corresponds to the standard dyadic intervals, with $[0, 1]$ being the root. Define the *left child* of $[a, b]$ to be $[a, (a + b)/2]$ and the *right child* to be $[(a+b)/2, b]$. The "top" of $\mathcal{T}$ is shown in Figure 10.1. Every finite, rooted binary tree $T$ can be viewed as a subtree of $\mathcal{T}$, where

---

[1] Recall that the dyadic rationals are all rational numbers of the form $\dfrac{m}{2^n}$ where $m$ and $n$ are integers. They are discussed in our presentation of BS$(1, 2)$ in Chapter 4.

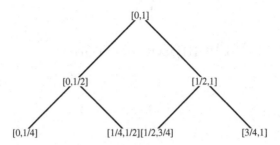

Fig. 10.1. The dyadic intervals generate a rooted binary tree.

the root of $T$ is identified with $[0, 1]$. The standard dyadic intervals that form the leaves of such a finite, rooted binary tree give a dyadic division of $[0, 1]$, and any dyadic subdivision generates a finite, rooted binary tree. As we will frequently refer to "finite, rooted binary trees" in the remainder of this chapter, we abbreviate our terminology to "frb-trees."

Given an ordered pair of dyadic divisions of $[0, 1]$, with the same number of pieces, there is a corresponding piecewise linear function $f : [0, 1] \to [0, 1]$. If $0 < m_1 < \cdots < m_k < 1$ denote the first set of chosen middles and $0 < \mu_1 < \cdots < \mu_k < 1$ denote the second set of chosen middles, then $f$ is defined by requiring:

1. $f(0) = 0$ and $f(1) = 1$;
2. $f(m_i) = \mu_i$ for all $i$;
3. $f$ is linear when restricted to $[0, m_1]$, $[m_i, m_{i+1}]$ (for $1 \le i \le k$) and $[m_k, 1]$.

For example, if the two dyadic divisions are

$$[0, 1/2] \cup [1/2, 3/4] \cup [3/4, 1]$$

and

$$[0, 1/4] \cup [1/4, 1/2] \cup [1/2, 1],$$

then the graph of the associated function is the one shown on the left in Figure 10.2. If the two divisions have chosen middles $\{1/2, 3/4, 7/8\}$ and $\{1/2, 5/8, 3/4\}$ then the associated function is the one on the right in Figure 10.2. We will refer to functions created in this manner as *Thompson functions*. We show in Proposition 10.3 that the set of all Thompson functions forms a group whose binary operation is function composition. This group is called *Thompson's group F*.

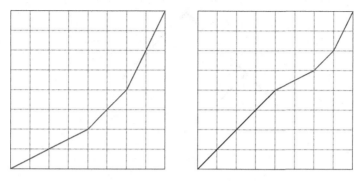

Fig. 10.2. Two elements of Thompson's group $F$.

Thompson functions are continuous bijections of $[0, 1]$, hence they can be viewed as elements of the group $\mathrm{Homeo}([0, 1])$, whose binary operation is function composition. Thus, to show that this subset is a subgroup, it suffices to show that it is closed under inverses and composition. To do this, we introduce a bit of notation. Since dyadic divisions of $[0, 1]$ correspond with frb-trees, we may use an ordered pair of frb-trees to describe elements of the group $F$. In particular, we will use $[T_2 \leftarrow T_1]$ to denote the Thompson function where the domain has been divided according to $T_1$ and the range according to $T_2$. The reader should verify the following lemma, which exploits this notation.

**Lemma 10.1.** *Let $T_1, T_2$ and $T_3$ be frb-trees with the same number of leaves. Then*

$$[T_3 \leftarrow T_2][T_2 \leftarrow T_1] = [T_3 \leftarrow T_1]$$

*and*

$$[T_2 \leftarrow T_1]^{-1} = [T_1 \leftarrow T_2].$$

Given an ordered pair of frb-trees (with the same number of leaves) there is an associated Thompson function. However, many different ordered pairs of frb-trees define the same element of $F$. For example, the identity corresponds to $[T \leftarrow T]$ for any frb-tree $T$. If $T$ is a frb-tree, let $T \wedge i$ denote the frb-tree created by adding a "wedge" or "caret" to $T$ at bottom of the $i$th leaf, where the leaves are enumerated left-to-right, with the leftmost leaf being numbered 0. An example is shown in Figure 10.3. Notice that the functions associated to the pair $[T_2 \leftarrow T_1]$ and to $[T_2 \wedge i \leftarrow T_1 \wedge i]$ are the same functions. Each has simply had their $(i + 1)$st dyadic interval divided in half, which does not alter the

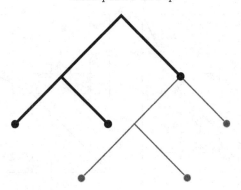

Fig. 10.3. A finite, rooted binary tree $T$ is shown above, in darker and thicker lines. The lighter edges show the result of forming $(T \wedge 2) \wedge 2$.

associated functions. Let $\sim$ be the equivalence relation generated by

$$[T_2 \leftarrow T_1] \sim [T_2 \wedge i \leftarrow T_1 \wedge i].$$

The functions defined by any two elements in an equivalence class are the same.

Since equivalence relations are symmetric, not only are pairs of frb-trees equivalent to pairs with carets added to their leaves, but sometimes pairs of carets can be deleted. If the leaves numbered $i$ and $i+1$ form a caret in $T$, then this is an *exposed caret*. This is equivalent to saying that there is an frb-tree $T'$ such that $T = T' \wedge i$. If $T_1$ and $T_2$ have matched, exposed carets, that is $T_1 = T_1' \wedge i$ and $T_2 = T_2' \wedge i$, then

$$[T_2' \leftarrow T_1'] \sim [T_2 \leftarrow T_1].$$

If $T_1$ and $T_2$ do not have a pair of matched, exposed carets, then the ordered pair $[T_2 \leftarrow T_1]$ is *reduced*. Exercise 3 asks you to prove the following helpful fact.

**Lemma 10.2.** *Each equivalence class contains a unique reduced pair of frb-trees.*

We have now introduced enough terminology to establish that we are indeed discussing a group.

**Proposition 10.3.** *The set of Thompson functions forms a group under function composition.*

*Proof.* All that remains to be shown is that the composition of Thompson functions is another Thompson function. Viewing frb-trees as subtrees of $\mathcal{T}$, you can form unions $T_1 \cup T_2$, and the result is another frb-tree. This process can be thought of in terms of overhead transparencies: take two transparencies, one with $T_1$ on it and the other containing $T_2$; lay one transparency over the other, lining up the roots, and you have produced $T_1 \cup T_2$. As each $T_i$ ($i = 1$ or 2) is a subtree of the union, each $T_i$ can be expanded to $T_1 \cup T_2$ by adding carets.

Given two pairs of frb-trees, $[T_2 \leftarrow T_1]$ and $[T_4 \leftarrow T_3]$, there is a set of equivalent pairs of frb-trees, $[T'_2 \leftarrow T'_1]$ and $[T'_4 \leftarrow T'_3]$, where $T'_2 = T'_3 = T_2 \cup T_3$. Thus, by Lemma 10.1, the composition of the associated functions is the function associated to $[T'_4 \leftarrow T'_1]$. Thus the set of Thompson functions is closed under composition; as we already knew they were closed under inversion, the set of Thompson functions forms a group. □

Directly from the definition we can derive a surprising fact about $F$. Define the *support* of $f \in F$ to be

$$\mathrm{Supp}(f) = \{x \in [0,1] \mid f(x) \neq x\}.$$

Call an element $f \in F$ a *left* element if $\mathrm{Supp}(f) \subset (0, 1/2)$. Similarly $f \in F$ is a *right* element if $\mathrm{Supp}(f) \subset (1/2, 1)$. The set of left elements forms a subgroup of $F$, which we denote $F_l$; the set of right elements forms a subgroup, $F_r$. Let $l$ be the homomorphism $F \to F$ that takes $f \in F$ to the function

$$f_l(x) = \begin{cases} f(2x)/2 & 0 \leq x \leq 1/2 \\ x & 1/2 \leq x \leq 1. \end{cases}$$

The graph of $f_l$ consists of a copy of the graph of $f$ that has been shrunk and tucked into $[0, 1/2] \times [0, 1/2]$, and then is extended by the identity on the remainder of the domain. Similarly let $r$ be the homomorphism $F \to F$ that takes $f \in F$ to the function

$$f_r(x) = \begin{cases} x & 0 \leq x \leq 1/2 \\ 1/2 + f(2x)/2 & 1/2 \leq x \leq 1. \end{cases}$$

The homomorphisms $l$ and $r$ show that both $F_l$ and $F_r$ are isomorphic to $F$. Since the support of elements in $F_l$ is disjoint from the support of elements in $F_r$, elements of $F_l$ commute with elements of $F_r$. Thus the subgroup generated by $F_l$ and $F_r$ is $F_l \oplus F_r \approx F \oplus F$. We have now outlined an argument that shows the following:

**Proposition 10.4.** *Thompson's group $F$ contains a subgroup isomorphic to $F \oplus F$.*

In order to construct a generating set for $F$, we introduced two infinite families of frb-trees. Let $\mathcal{T}_n$ be the frb-tree where $\mathcal{T}_0$ is a single caret and $\mathcal{T}_{n+1} = \mathcal{T}_n \wedge (n+1)$. Let $\mathcal{S}_n$ be the frb-tree where $\mathcal{S}_0 = \mathcal{T}_0$ and $\mathcal{S}_{n+1} = \mathcal{T}_n \wedge n$. (Examples are shown in Figure 10.4.)

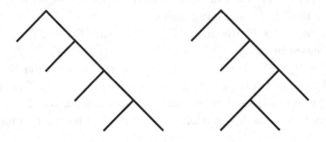

Fig. 10.4. The frb-trees $\mathcal{T}_3$ and $\mathcal{S}_3$.

An element $f \in F$ is said to be *positive* if it corresponds to an ordered pair of the form $[T \leftarrow \mathcal{T}_n]$; it is negative if it corresponds to an ordered pair $[\mathcal{T}_n \leftarrow T]$. Since $[\mathcal{T}_n \leftarrow T] = [T \leftarrow \mathcal{T}_n]^{-1}$, the negative elements are the inverses of the positive elements.

**Lemma 10.5.** *Every element of Thompson's group $F$ can be expressed as a product $p{\cdot}n$ where $p$ is a positive element and $n$ is a negative element of $F$.*

*Proof.* Let $f$ be the Thompson function given by $[S \leftarrow T]$ where $S$ and $T$ have $n+1$ leaves. Then

$$f = [S \leftarrow T] = [S \leftarrow \mathcal{T}_n][\mathcal{T}_n \leftarrow T]$$

expresses $f$ as a product of a positive and a negative element. $\square$

Define $x_i$ to be the Thompson function described by the ordered pair $[\mathcal{S}_{i+1} \leftarrow \mathcal{T}_{i+1}]$. The graphs of the elements $x_0$ and $x_1$ are shown in Figure 10.2, with $x_0$ on the left and $x_1$ on the right. The reader should note that $x_1$ is essentially just a copy of the function $x_0$, tucked into $[1/2, 1] \times [1/2, 1]$. In fact, the element $x_n$ is a rescaled copy of $x_0$ tucked into $[1 - (1/2)^n, 1] \times [1 - (1/2)^n, 1]$.

This collection of elements satisfies a number of nice relations, two of which we highlight in the following lemmas.

**Lemma 10.6.** *If $i < n$ then $x_n x_i = x_i x_{n+1}$.*

This relation can be rewritten as: $x_i^{-1} x_n x_i = x_{n+1}$. As the function $x_i$

takes the interval $[1-(1/2)^{i+2}, 1]$ linearly to $[1-(1/2)^{i+1}, 1]$, Lemma 10.6 is an application of the principle that "conjugation is doing the same thing somewhere else."

**Lemma 10.7.** *Let* $i < n + 2$. *Then* $[T \leftarrow \mathcal{T}_n] \cdot x_i = [T \wedge i \leftarrow \mathcal{T}_{n+1}]$.

*Proof.* By definition the element $x_i$ is $[\mathcal{S}_{i+1} \leftarrow \mathcal{T}_{i+1}]$. Since $i < n + 2$, $\mathcal{T}_n \cup \mathcal{S}_{i+1} = \mathcal{T}_n \wedge i$. Thus we have

$$\begin{aligned}
[T \leftarrow \mathcal{T}_n] \cdot x_i &= [T \leftarrow \mathcal{T}_n] \cdot [\mathcal{S}_{i+1} \leftarrow \mathcal{T}_{i+1}] \\
&= [T \leftarrow \mathcal{T}_n] \cdot [\mathcal{T}_n \wedge i \leftarrow \mathcal{T}_{n+1}] \\
&= [T \wedge i \leftarrow \mathcal{T}_n \wedge i] \cdot [\mathcal{T}_n \wedge i \leftarrow \mathcal{T}_{n+1}] \\
&= [T \wedge i \leftarrow \mathcal{T}_{n+1}]
\end{aligned}$$

$\square$

**Theorem 10.8.** *Thompson's group $F$ is generated by the infinite set of positive elements* $\{x_0, x_1, x_2, \dots\}$.

*Proof.* As every element of $F$ is a product of a positive and a negative element (Lemma 10.5), and the negative elements are inverses of positive elements, it suffices to show that every positive element of $F$ is a product of the $x_i$'s. Let $[T \leftarrow \mathcal{T}_n]$ be a positive element. Note that $T$ has a maximal subtree of the form $\mathcal{T}_k$, such that $T = \mathcal{T}_k \wedge i_1 \wedge i_2 \wedge \cdots \wedge i_m$, where one never adds a caret to the rightmost leaf. The formula given in Lemma 10.7 then shows

$$[T \leftarrow \mathcal{T}_n] = [\mathcal{T}_k \leftarrow \mathcal{T}_k] \cdot x_{i_1} \cdots x_{i_m}.$$

Since $[\mathcal{T}_k \leftarrow \mathcal{T}_k]$ is just the identity element of $F$, it follows that the positive element $[T \leftarrow \mathcal{T}_n]$ is $x_{i_1} \cdots x_{i_m}$. $\square$

**Corollary 10.9.** *Thompson's group $F$ is generated by two elements, $x_0$ and $x_1$.*

*Proof.* Lemma 10.6 shows that $x_2 = x_0^{-1} x_1 x_0$. Thus an induction argument shows $x_{n+1} = x_0^{-n} x_1 x_0^n$. Since every element of $F$ can be expressed as a product of the $x_i$ and their inverses, and every $x_i$ can be expressed as a word in the alphabet $\{x_0, x_1, x_0^{-1}, x_1^{-1}\}$, the result follows. $\square$

We conclude by quoting a few known results about Thompson's group $F$. Proofs of the first two results, along with a number of other interesting facts, can be found in the survey [CFP96].

**Theorem 10.10.** *Every proper quotient of Thompson's group $F$ is abelian.*

**Theorem 10.11.** *Every non-abelian subgroup of F contains a copy of the wreath product $\mathbb{Z} \wr \mathbb{Z}$. In particular, since every subgroup of a free group is free, Thompson's group F does not contain a subgroup isomorphic to the free group $\mathbb{F}_2$.*

The next set of results relate to topics introduced in Chapter 9.

**Theorem 10.12** ([GuSa97]). *Thompson's group F admits a regular normal form.*

The normal form constructed by Guba and Sapir in [GuSa97] is easy to describe. It consists of all words in $\{x_0, x_1, x_0^{-1}, x_1^{-1}\}^*$ which do not contain one of these elements followed by its inverse, and which do not contain any subword of the form

1. $x_1 x_0^i x_1$,
2. $x_1^{-1} x_0^i x_1$,
3. $x_1 x_0^{i+1} x_1^{-1}$,
4. $x_1^{-1} x_0^{i+1} x_1^{-1}$.

where $i$ is any positive integer. Checking that this set of words is a regular language is routine; proving that it is a normal form is not. It should be noted that this normal form is not composed of geodesic words. This was not due to a lack of insight on the part of the authors. Rather, it was later shown by Cleary, Elder and Taback that it is impossible to construct such a normal form.

**Theorem 10.13** ([CET06]). *The Cayley graph of Thompson's group F with respect to $\{x_0, x_1\}$ has infinitely many cone types. In particular, there is no regular normal form consisting of geodesic words.*

This shows that one cannot hope to show that the growth series for $F$ is rational using the approach of Section 9.3. As of the time of this writing, it is unknown if the growth series for Thompson's group $F$ is rational or not.

Finally, the geometry of the Cayley graph of $F$, with respect to $\{x_0, x_1\}$, is quite complicated. In particular, Belk and Bux have shown:

**Theorem 10.14** ([BeBu05]). *The Cayley graph of F with respect to $\{x_0, x_1\}$ is not minimally almost convex.*

This work exploits a method introduced by Blake Fordham in his PhD thesis that gives a simple method of computing the length of elements in $F$ with respect to the generators $\{x_0, x_1\}$. An alternative, also elementary, method for computing length was later introduced by Victor Guba. (See [BeBr05] and [Gu04] for reasonably accessible discussions of these methods.) It is a rather annoying state of affairs that elementary and elegant means of computing length are available, but many questions about the geometry of the Cayley graph of $F$ with respect to $\{x_0, x_1\}$ remain unresolved.

**Remark 10.15.** Thompson's group $F$ was introduced in 1965 by Richard Thompson. His original treatment of the group was motivated by algebraic questions arising from formal logic. Since then this group has appeared in many different contexts, from deep studies of the associative law to topological considerations of idempotents.

In addition to $F$, there are also Thompson groups $T$ and $V$ that add permutations into the mix. Similar groups where the slopes are not as constricted, for instance allowing slopes of the form $2^m 3^n$, have also been studied. The previously mentioned survey [CFP96] is a well-written, fairly elementary introduction to the groups $F, T$ and $V$, as well as some of the known variations.

## Exercises

(1) Show that if you divide $[0, 1]$ into dyadic intervals, then this subdivision must be a dyadic subdivision.

(2) Prove that a continuous, bijective, piecewise linear function $f : [0, 1] \to [0, 1]$ (with finitely many corners) is a Thompson function if and only if:

    a. all of the slopes (away from the corners) are powers of 2; and

    b. all of the corners have both coordinates in the dyadic rationals.

(3) Prove Lemma 10.2.

(4) Prove that $F$ is torsion-free.

(5) Let $\mathcal{I} = \left[\dfrac{m}{2^n}, \dfrac{m+1}{2^n}\right]$ be a standard dyadic interval. Let $F_{\mathcal{I}}$ be the subset of functions

$$F_{\mathcal{I}} = \{f \in F \mid \mathrm{Supp}(f) \subset \mathcal{I}\}.$$

    a. Prove that $F_{\mathcal{I}}$ is a subgroup.

    b. Prove that $F_{\mathcal{I}} \approx F$.

(6) Show that the set of positive elements of $F$ is closed under composition.

(7) In this problem we discuss the standard normal form for elements of Thompson's group $F$ using $\{x_0, x_1, x_2, \ldots\}$.

    a. Show that every element of $F$ can be expressed as a word of the form

$$x_0^{a_0} x_1^{a_1} \cdots x_n^{a_n} x_n^{-b_n} \cdots x_1^{-b_1} x_0^{-b_0}$$

    where the $a_i$ and $b_i$ are non-negative integers. We refer to such expressions as *up-and-down* words.

    b. For an frb-tree $T$, define the *exponent* of a leaf to be the length of the maximal path consisting only of left edges, that contains the leaf but does not reach the right side of $T$. (An example is shown in Figure 10.5.) If $\{a_0, a_1, \ldots, a_m\}$

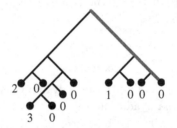

Fig. 10.5. The exponents of the leaves in this frb-tree, $T$, are as indicated. This implies the element $[T \leftarrow T_8]$ can be expressed as $x_0^2 x_2^3 x_6$.

    are the exponents of the leaves of an frb-tree $T$, show that the positive element $[T \leftarrow T_{m-1}]$ is $x_0^{a_0} x_1^{a_1} \cdots x_m^{a_m}$.

    c. If $f \in F$ corresponds to $[S \leftarrow T]$ describe how to express $f$ as an up-and-down word using the exponents of $S$ and $T$.

    d. Show that $[S \leftarrow T]$ is a reduced diagram if and only if its associated up-and-down word $x_0^{a_0} \cdots x_n^{a_n} x_n^{-b_n} \cdots x_0^{-b_0}$ satisfies:

        1. exactly one of $a_n$ and $b_n$ is non-zero; and

        2. if $a_i$ and $b_i$ are both non-zero, then either $a_{i+1} \neq 0$ or $b_{i+1} \neq 0$.

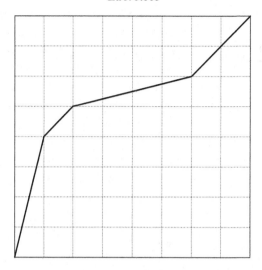

Fig. 10.6. An element of Thompson's group $F$.

(8) An element of $F$ is shown in Figure 10.6. Express this element as a word in $\{x_0, x_1, x_0^{-1}, x_1^{-1}\}^*$.

# 11

# The Large-Scale Geometry of Groups

What one really cares about are the inherent properties of the group, not the artefacts of a particular presentation.

–Martin Bridson

## 11.1 Changing Generators

There is a danger in working with a specific Cayley graph for a given group $G$. If you focus on a particular generating set, the results you get may not immediately translate into similar results when a different set of generators is used. Even worse, sometimes interesting properties hold in one Cayley graph but not in another, even though the group under consideration has not changed.

In this chapter we introduce some of the ways geometry can be imported into the study of infinite groups, which are independent of the choice of finite generating set. These are properties that always hold or always fail, no matter which finite set of generators one uses to construct a Cayley graph. Of course, such properties cannot focus too closely on a given Cayley graph, since changing generators changes the details, in other words the local structure, of the graphs. In Figure 11.1 we highlight this with two Cayley graphs for $D_3$. The location of the vertices has been kept constant in both pictures, which makes the second picture look a bit odd. When the generators change, so do the distances between vertices, the number of cycles, and so on.

Because changing generators can have dramatic consequences for the local structure of a Cayley graph, the properties we consider here are referred to as *large-scale properties*; many authors also use the term *geometric properties*. A property of a Cayley graph of a finitely generated group is large-scale, or geometric, only if it is invariant under changes in

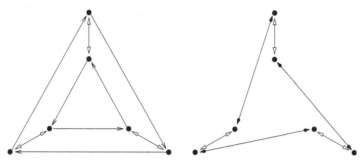

Fig. 11.1. The Cayley graph of $D_3$ with respect to two different generating sets. The Cayley graph on the left corresponds to using a reflection and a rotation; the Cayley graph on the right corresponds to using two reflections.

generating set.[1] With this convention, the property of having a rational growth series is not a geometric property of a group (see Remark 9.18). On the other hand, there is a notion of growth that is geometric, which we discuss in Section 11.2.

In this section we introduce a standard technique for passing between Cayley graphs of a fixed group $G$.

**Definition 11.1.** Let $\Gamma$ and $\Lambda$ be two graphs. A *map* from $\Gamma$ to $\Lambda$ is a function $\phi$ taking vertices of $\Gamma$ to vertices of $\Lambda$, and edges of $\Gamma$ to edge paths in $\Lambda$, such that if $v$ and $w$ are vertices attached to an edge $e \in \Gamma$, then $\phi(e)$ joins $\phi(v)$ to $\phi(w)$.

As an example, consider a tree $T$ consisting of three vertices and two edges, along with the Cayley graph $\Gamma$ of $\mathbb{Z}$ with respect to $\{2, 3\}$. Send the vertices of $T$ to the vertices corresponding to $0, 1$ and $2$ in $\Gamma$, and map each edge of $T$ to the edge path corresponding to $\{3, -2\}$. The result is indicated in Figure 11.2.

**Proposition 11.2.** *Let $S$ and $T$ be two finite generating sets for a group $G$ and let $\Gamma_S$ and $\Gamma_T$ be the corresponding Cayley graphs. Then there are maps $\phi_{T \leftarrow S} : \Gamma_S \to \Gamma_T$ and $\phi_{S \leftarrow T} : \Gamma_T \to \Gamma_S$ such that:*

1. *The compositions $\phi_{S \leftarrow T} \circ \phi_{T \leftarrow S}$ and $\phi_{T \leftarrow S} \circ \phi_{S \leftarrow T}$ induce the identity on $V(\Gamma_S)$ and $V(\Gamma_T)$ respectively.*
2. *There is a constant $K > 0$ such that the image of any edge $e \in \Gamma_S$*

---

[1] Actually, the term is used more broadly than this; it is not restricted to Cayley graphs or to changes in generating sets. In particular, these terms are used in the study of metric spaces where a property is *geometric* if it is invariant under "quasi-isometry." This is a term we introduce in Section 11.7.

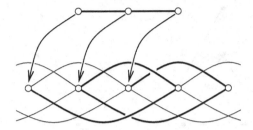

Fig. 11.2. A map from a finite tree to the Cayley graph of $\mathbb{Z}$ with respect to the generating set $\{2,3\}$.

*under $\phi_{S\leftarrow T} \circ \phi_{T\leftarrow S}$ is contained in the ball $\mathcal{B}(v, K) \subset \Gamma_S$, where $v \in \mathrm{ENDS}(e)$. The similar statement also holds for edges of $\Gamma_T$.*

*Proof.* The map $\phi_{T\leftarrow S}$ is the identity map on the vertices. That is, if $v_g$ denotes the vertex corresponding to $g \in G$ in $\Gamma_S$, and $v'_g$ denotes the vertex corresponding to $g$ in $\Gamma_T$, then $\phi_{T\leftarrow S}(v_g) = v'_g$. The map $\phi_{S\leftarrow T}$ is defined in the same way on vertices, and the first claim is then immediate.

For each generator $s \in S$ choose a word $\omega_s = t_1 t_2 \cdots t_k \in \{T \cup T^{-1}\}^*$ such that $s = \pi(w_s) \in G$. (One can always do this since $T$ is a generating set for $G$.) By the construction of Cayley graphs, if $e$ is an edge of $\Gamma_S$ then $e$ is labelled by a generator $s \in S$ and joints the vertex associated to $g$ to the vertex associated to $g \cdot s$. The map $\phi_{T\leftarrow S}$ sends every such edge to the edge path

$$g \longrightarrow g \cdot t_1 \longrightarrow g t_1 \cdot t_2 \longrightarrow \cdots \longrightarrow g t_1 t_2 \cdots t_k$$

in $\Gamma_T$.

The map $\phi_{S\leftarrow T}$ is defined similarly. For each generator $t \in T$ choose a word $\omega_t \in \{S \cup S^{-1}\}^*$ representing $t$ and use these words to describe the images of the edges of $\Gamma_T$.

Let $k$ be the maximal length of the words $\omega_t$ and $\omega_s$. It follows that $\phi_{T\leftarrow S}(e)$ is an edge path of length $\leq k$. The image $\phi_{S\leftarrow T}$ of this path is then an edge path of length $\leq k^2$. Hence the constant $K$ in the second claim of the theorem can be taken to be $k^2$. ☐

**Corollary 11.3.** *Let $G, S$ and $T$ be as above. Then there is a constant $\lambda \geq 1$ such that for any $g$ and $h$ in $G$,*

$$\frac{1}{\lambda} d_S(g,h) \leq d_T(g,h) \leq \lambda d_S(g,h).$$

*Proof.* Let

$$\Lambda_1 = \mathrm{Max}\{|\omega_t| \mid t \in T\}$$

be the maximum length of the words in $\{S \cup S^{-1}\}^*$ which were chosen to represent the generators in $T$. If $d_T(g,h) = n$, then there is an edge path between $v_g$ and $v_h$ in $\Gamma_T$ of length $n$. This path maps under $\phi_{S \leftarrow T}$ to an edge path joining $v_g$ and $v_h$ in $\Gamma_S$ whose length is at most $\Lambda_1 \cdot n$. Thus $d_S(g,h) \le \Lambda_1 \cdot d_T(g,h)$. Repeating this argument, with the roles of $S$ and $T$ reversed, shows that if $\Lambda_2 = \mathrm{Max}\{|\omega_s| \mid s \in S\}$, then $d_T(g,h) \le \Lambda_2 \cdot d_S(g,h)$. Set $\lambda = \mathrm{Max}(\Lambda_1, \Lambda_2)$, and the stated inequalities hold. $\square$

**Example 11.4.** In Chapter 2 we introduced the infinite dihedral group $D_\infty$, which is generated by two reflections $a$ and $b$. The product $\tau = ab$ is a translation and $\{a, \tau\}$ is an alternate generating set for $D_\infty$. The Cayley graph of $D_\infty$ with respect to $S = \{a, b\}$ is shown in Figure 2.2 and the Cayley graph with respect to $T = \{a, \tau\}$ is shown below in Figure 11.3. If $\rho$ is a rotation in a finite dihedral group, and $a$ is a reflection, then $a\rho = \rho^{-1}a$. Similarly in the infinite dihedral group we have $a\tau = \tau^{-1}a$, as can be seen in the Cayley graph.

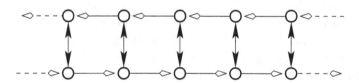

Fig. 11.3. The Cayley graph of $D_\infty$ with respect to the generating set $\{a, \tau\}$.

We may express the original generators in terms of the new generators by setting $\omega_a = a$ and $\omega_b = a\tau$. We can express the new generators in terms of the old generators by setting $\omega_a = a$ (again) and $\omega_\tau = ab$. The map $\phi_{T \leftarrow S}$ will send edges labelled $b$ to edge paths labelled $a\tau$, and $\phi_{S \leftarrow T}$ will send edges labelled $\tau$ to edge paths labelled $ab$. The constant $\lambda$ from Corollary 11.3 can be taken to be 2. Thus in passing from the original generating set to the new generating set (or vice versa) the distance between vertices is altered by at most a factor of 2.

In this example you can begin to see, in a very vague sense, that some sort of geometric structure is preserved when you switch generating sets. The Cayley graphs $\Gamma_S$ and $\Gamma_T$ are different. Yet, if one were to step well back from their graphs, in both instances it would seem as if you are looking at the real line.

**Remark 11.5.** In Chapter 5 we discussed Dehn's word problem. There are a number of geometric approaches to the word problem, but all of them require some familiarity with fundamental groups and covering spaces, so they will not be discussed in this book. The reader with an interest in this topic, who has the appropriate background, should take a look at [BRS07] and the references cited there.

## 11.2 The Growth of Groups, II

In this section we revisit the notion of growth in groups, first discussed in Section 9.2. Part of our discussion applies more generally to growth in metric spaces, and specifically the notion of growth functions. Define any non-decreasing function $f : [0, \infty) \to [0, \infty)$ to be a *growth function*. For example, we could take $f(x)$ to be the area enclosed by a circle of radius $x$ in the Euclidean plane, in which case $f(x) = \pi x^2$.

Our interest is in measuring the growth of groups. Let $G$ be a group with finite generating set $S$, and let $\mathcal{B}_S(n)$ denote the ball of radius $n$ about the identity in $G$. (This time we are thinking of $\mathcal{B}_S(n)$ as a subset of $G$, not a subgraph of $\Gamma_{G,S}$.) The function $\beta_S(n) = |\mathcal{B}_S(n)|$ gives the number of elements in $G$ inside the ball of radius $n$, that is, the number of elements of $G$ that can be expressed as a product of at most $n$ generators, and their inverses, coming from $S$. We extend the domain of such growth functions by defining $\beta(x)$ to be $\beta(\lfloor x \rfloor)$, for all $x \in [0, \infty)$. (Although when we state explicit formulas we do so viewing $\beta_S$ as a function from $\mathbb{N}$ to $\mathbb{N}$.)

Different generating sets for a fixed group $G$ do produce different growth functions. For example, if we take two reflections as our generating set for $D_\infty$, $S = \{a, b\}$, then $\beta_S(n) = 2n + 1$. On the other hand, we may generate $D_\infty$ by $\{a, \tau\}$ where $\tau = ab$ is a translation. The Cayley graph with respect to this generating set is shown in Figure 11.3. Using this figure, you will find that $\beta_T(n) = 4n$ (for $n \geq 1$). While $\beta_S$ and $\beta_T$ are not the same, both functions are linear, and it can be proved that this fact is independent of the generators one chooses. This is the idea we explore in this section: The rate at which a group expands (linearly, quadratically, ..., exponentially) is a geometric property of the group.

**Lemma 11.6.** *Let $S$ and $T$ be two finite generating sets for a fixed group $G$. Then there is a constant $\lambda \geq 1$ such that*

$$\beta_S\left(\frac{1}{\lambda}n\right) \leq \beta_T(n) \leq \beta_S(\lambda n)$$

*for all $n \in \mathbb{N}$.*

*Proof.* Let $|g|_S$ denote the length of $g \in G$ with respect to the generating set $S$, and define $|g|_T$ similarly. By Corollary 11.3, there is a $\lambda \geq 1$ such that $|g|_T \leq \lambda \cdot |g|_S$ for all $g \in G$. Thus if $g \in \mathcal{B}_T(n)$ then $g \in \mathcal{B}_S(\lambda n)$, which implies $\beta_T(n) \leq \beta_S(\lambda n)$. Exchanging the roles of $S$ and $T$ establishes the other inequality. □

Lemma 11.6 is one motivation for the following definition; another motivation arises in Section 11.7.

**Definition 11.7.** Define $\preceq$ to be the relation on growth functions defined by $f \preceq g$ if there is a constant $\lambda \geq 1$ such that

$$f(x) \leq \lambda g(\lambda x + \lambda) + \lambda$$

for all $x \in [0, \infty)$. If $f \preceq g$ then we say $g$ *dominates* $f$. (Notice that the right-hand side is the result of pre- and post-composing $g(x)$ with the linear function $y = \lambda x + \lambda$.)

If $f \preceq g$ and $g \not\preceq f$ then $g$ *strictly dominates* $f$. We denote this by $f \prec g$. On the other hand, if $f \preceq g$ and $g \preceq f$ then $f$ and $g$ are said to be *equivalent*, denoted $f \sim g$. Two equivalent growth functions are said to *grow at the same rate*.

Exercise 2 at the end of this chapter asks you to prove that "$\preceq$" is a reflexive and transitive relation, and that $f \sim g$ is an equivalence relation.

**Example 11.8.** Consider the following growth functions: $f(n) = n^2$, $g(n) = 15n^2 + 10n + 1$ and $h(n) = 2^n$. We have $f(n) < g(n)$ for all $n \in \mathbb{N}$, so $f \preceq g$. On the other hand,

$$f(5n) + 5 = 25n^2 + 5 = 15n^2 + 10n^2 + 5 > 15n^2 + 10n + 1 = g(n)$$

so $g \preceq f$, hence $f \sim g$.

One expects that $f \preceq h$, since exponential functions grow faster than polynomial functions. There is a slight hitch in that $f(3) = 9 > 8 = h(3)$ but, for all other $n \in \mathbb{N}$, $f(n) \leq h(n)$. Thus $f(n) < h(2n) + 2$ for all $n$ and so it is indeed the case that $f \preceq h$. On the other hand,

$$\lim_{n \to \infty} \frac{\lambda(\lambda n + \lambda)^2 + \lambda}{2^n} = 0$$

for any $\lambda \geq 1$. Thus $f \not\preceq h$ and so $h$ strictly dominates $f$.

Lemma 11.6 implies the following:

**Corollary 11.9.** *If $S$ and $T$ are two finite generating sets for a group $G$ then the associated growth functions are equivalent:* $\beta_S(n) \sim \beta_T(n)$.

The equivalence class of a growth function for a group $G$ is then a large-scale invariant of the group. For example, if $\beta(n)$ is a growth function for $D_\infty$, then $\beta(n) \sim n$. Hence one one can say that the infinite dihedral group has *linear growth*.

**Definition 11.10.** Let $G$ be some fixed group with $\beta(n)$ an associated growth function. Then $G$ is said to have *polynomial growth* if there is some $d \in \mathbb{N}$ such that $\beta(n) \preceq n^d$. The group $G$ has *polynomial growth of degree $d$* if $\beta(n) \sim n^d$.

It is easy to check that the growth function for $\mathbb{Z}$ is $\beta(n) = 2n + 1$, with respect to a cyclic generator. Hence $\mathbb{Z}$ is another example of a group with linear growth. You can also establish that $\beta(n) = 2n^2 + 2n + 1$ for $\mathbb{Z} \oplus \mathbb{Z}$ with respect to its standard generating set. Thus $\mathbb{Z}^2$ is said to have quadratic growth. These examples should lead us to conjecture that free abelian groups of rank $d$ have polynomial growth of degree $d$. One could verify this by explicitly computing formulas for each $\beta(n)$, with respect to the standard generators of $\mathbb{Z}^d$. But when looking at large-scale issues one rarely needs such precise information. The elements in $\mathbb{Z}^d$ that are contained in the ball of radius $n$ consist of those $d$-tuples of integers $(n_1, n_2, \ldots, n_d)$ where $|n_1| + |n_2| + \cdots + |n_d| \leq n$. A standard combinatorial argument shows there are $\binom{n+d-1}{d}$ such $d$-tuples whose entries are non-negative. Thus $\binom{n+d-1}{d} \preceq \beta(n)$. Similarly, if we allow both positive and negative entries, you can get the bound $\beta(n) \preceq 2^d \cdot \binom{n+d-1}{d}$. Thus the growth function for $\mathbb{Z}^d$, with respect to the standard generators, satisfies $\beta(n) \sim \binom{n+d-1}{d} \sim n^d$. We therefore have:

**Proposition 11.11.** *The free abelian group* $\mathbb{Z}^d = \underbrace{\mathbb{Z} \oplus \cdots \oplus \mathbb{Z}}_{d \text{ copies}}$ *has polynomial growth of degree $d$.*

Not all growth functions of groups are polynomial. Consider the free group of rank $k$, $\mathbb{F}_k$, along with a fixed basis. The number of elements in the sphere of radius $n$, $\mathcal{S}(n)$, is $2k \cdot (2k-1)^{n-1}$ (for $n \geq 1$). Thus $(2k-1)^n < |\mathcal{S}(n)| < |\mathcal{B}(n)|$. Hence the growth function of $\mathbb{F}_k$ satisfies $(2k-1)^n \preceq \beta(n)$. In particular, as long as $k \geq 2$, the growth is *at least* exponential. On the other hand,

$$\beta(n) = \sum_{i=0}^{n} |\mathcal{S}(n)| = 1 + 2k + 2k \cdot (2k - 1) + \cdots + 2k \cdot (2k - 1)^{n-1}$$
$$< 1 + 2k + (2k)^2 + \cdots (2k)^{n-1} < (2k)^n.$$

So $\beta(n) \preceq (2k)^n$, hence the growth is *at most* exponential. But these bounding exponential functions are equivalent.

**Lemma 11.12.** *Let $a$ and $b$ be two integers, both greater than 1, and let $\alpha(n) = a^n$ and $\beta(n) = b^n$ be the corresponding exponential functions. Then $\alpha(n) \sim \beta(n)$.*

*Proof.* We will assume that $a < b$ so that $\alpha(n) \preceq \beta(n)$ trivially. Conversely, if $\lambda \geq \log_a(b)$, then $\beta(n) \leq \alpha(\lambda n) + \lambda$, so $\beta(n) \preceq \alpha(n)$. □

A group $G$ is said to have *exponential growth* if its associated growth function satisfies $\beta(n) \sim 2^n$. Lemma 11.12 shows that the base 2 could be replaced by any integer $\geq 2$, and this combined with the previous discussion of free groups shows:

**Proposition 11.13.** *Every finitely generated free group, whose rank is at least 2, has exponential growth. Further, if $\beta_n$ represents a growth function for $\mathbb{F}_n$, then $\beta_n \sim \beta_m$ for any integers $m$ and $n$ that are both greater than or equal to 2.*

**Corollary 11.14.** *If $G$ is a finitely generated group, then its growth is dominated by $2^n$.*

*Proof.* Assume $G$ has a generating set with $k$ elements. As every element of $G$ can be expressed as a freely reduced word in its generators and their inverses, the associated growth function is bounded above by the growth function of $\mathbb{F}_k$. But the growth functions of free groups are either linear ($k = 1$) or equivalent to $2^n$ ($k > 1$). □

## 11.3 The Growth of Thompson's Group

In 1968 John Milnor asked if there are groups of "intermediate growth," which is to say: can the growth function of a finitely generated group $G$ strictly dominate $n^d$ for all $d$, and yet not be equivalent to $2^n$? There are functions with this property, such as $2^{\sqrt{n}}$, but can such functions arise as the growth function of a finitely generated group? Since Thompson's group $F$ contains a copy of $F \oplus F$ (Proposition 10.4), it also contains a copy of $F \oplus F \oplus F$, and so on. Thus Thompson's group $F$ contains a copy of $\mathbb{Z}^d$ for all $d$. So it is reasonable guess that the growth of $F$ dominates $n^d$ for all $d$. Further, $F$ does not contain a non-abelian free group (Theorem 10.11), so *perhaps* the growth of $F$ is not exponential.

**Theorem 11.15.** *Thompson's group $F$ has exponential growth.*

*Proof.* Recall that Thompson's group $F$ is generated by two elements, $x_0$ and $x_1$, whose graphs are shown in Figure 10.2. Let $\omega_0$ and $\omega_1$ be two words in $\{x_0, x_1\}^*$. (We are ignoring all inverses!) Our goal in this proof is to show that no two such words represent the same element of $F$, that is, $\pi(\omega_0) \neq \pi(\omega_1)$, where $\pi : \{x_0, x_1, x_0^{-1}, x_1^{-1}\}^* \to F$ is the evaluation map.

Assume to the contrary that $\pi(\omega_0) = \pi(\omega_1)$ for some pair of words in the generators $x_0$ and $x_1$. Choose an example where the combined length of $\omega_0$ and $\omega_1$ is as short as possible. Note that the last letters of such a pair must be distinct. If they both end in the same letter, "$x$," this letter can be removed from both words, forming $\omega_0'$ and $\omega_1'$. But

$$\begin{aligned}\pi(\omega_0') &= \pi(\omega_0 \cdot x^{-1}) \\ &= \pi(\omega_0)\pi(x^{-1}) \\ &= \pi(\omega_1)\pi(x^{-1}) \\ &= \pi(\omega_1').\end{aligned}$$

Thus the resulting pair would still satisfy $\pi(\omega_0') = \pi(\omega_1')$, but their combined length will have been reduced. Since $|\omega_0| + |\omega_1|$ was minimal, this is not possible, and so the two words end in distinct generators. Without loss of generality we assume $\omega_0$ ends in $x_0$ and $\omega_1$ ends in $x_1$.

If $f$ is an element of $F$, then when restricted to some small interval $[0, \epsilon)$, $f$ is linear with slope $2^k$. The function $\phi : F \to \mathbb{Z}$ that takes $f$ to this exponent $k$ is a homomorphism. (If you compose a function with slope $2^k$ with a function of slope $2^l$ then the result has slope $2^{k+l}$.) In particular, $\phi(x_0) = -1$ and $\phi(x_1) = 0$. Since $\omega_0$ and $\omega_1$ represent the same element of $F$, $\phi(\omega_0) = \phi(\omega_1)$. Further, $|\phi(\omega_0)| = |\phi(\omega_1)|$ is the number of $x_0$'s in these words. Call this number $n$. Note that $n > 0$ because otherwise $\omega_0$ and $\omega_1$ would simply be powers of $x_1$.

Thinking of $x_0$ as a function from $[0, 1]$ to $[0, 1]$ we see $x_0(3/4) = 1/2$ and, more generally, $x_0^k(3/4) = 1/2^k$ for any positive integer $k$. Since $x_1$ is the identity when restricted to $[0, 1/2]$, and $\omega_0$ ends in $x_0$, it follows that $\pi(\omega_0)$ takes $3/4$ to $1/2^n$ (where $n$ is as defined above).

Now consider the action of $\pi(\omega_1)$ on $[0,1]$. There is some positive $k$ such that $\omega_1$ ends with $x_0 x_1^k$. The function $x_1^k$ takes $3/4$ to some number strictly smaller than $3/4$, hence $x_0 x_1^k$ takes $3/4$ to some number strictly smaller than $1/2$. Any additional $x_1$'s that appear in $\omega_1$ have no effect on $[0, 1/2]$, but there are $n-1$ more $x_0$'s in $\omega_1$. Thus, in the end, $\pi(\omega_1)$ takes $3/4$ to some number which is strictly less than $1/2^n$. So it is impossible for $\omega_0$ and $\omega_1$ to represent the same element of Thompson's group $F$.

Because there are $2^n$ words of length $n$ in $\{x_0, x_1\}^*$, it follows that $2^n \leq \beta(n)$, and so Thompson's group $F$ has exponential growth. $\qquad\square$

While it is interesting to see that Thompson's group $F$ has exponential growth, and that this can be determined even though details about the growth series of $F$ are currently unknown, this example does not resolve Milnor's Question: are there groups of intermediate growth? In 1983, Grigorchuck proved the following:

**Theorem 11.16.** *There are finitely generated groups $G$ with growth function $\beta$ where*

$$n^d \prec \beta \prec 2^n$$

*for all $d$.*

The first known example of a group of intermediate growth was very similar to the infinite torsion group that we constructed in Chapter 6. A nice, accessible discussion of this result can be found in [GrPa07].

We close this section by quoting another important result, a full discussion of which is beyond the scope of this text. The notion of groups of polynomial growth is a geometric notion, defined in terms of metric properties of the group. Seemingly unrelated is the algebraic property of being nilpotent. Let $Z_1 = Z(G)$ be the center of $G$ and define $Z_{n+1} = \{g \in G \mid \text{for all } h \in G, hgh^{-1}g^{-1} \in Z_n\}$. Thus $Z_2/Z_1$ is the center of $G/Z_1$ and in general $Z_{n+1}/Z_n$ is the center of $G/Z_n$. A group $G$ is *nilpotent* if $G = Z_n$ for some $n$. Abelian groups are nilpotent since $G = Z(G) = Z_1$. The Heisenberg group $H$ consists of 3-by-3 integer matrices of the form:

$$\left\{ \begin{bmatrix} 1 & a & b \\ 0 & 1 & c \\ 0 & 0 & 1 \end{bmatrix} \mid a, b, c \in \mathbb{Z} \right\}.$$

The Heisenberg group $H$ is not abelian. However, $Z(H) \approx \mathbb{Z}$ and $H/Z(H) \approx \mathbb{Z} \oplus \mathbb{Z}$, so $H$ is nilpotent.

In 1981, Gromov published the following amazing result:

**Theorem 11.17.** *A finitely generated group has polynomial growth if and only if it is virtually nilpotent.*

The original paper, [Gr81], not only established this result, it also introduced advanced techniques that have proved to be quite useful in studying the geometry of infinite groups.

## 11.4 The Ends of Groups

Let $\Gamma$ be a connected, locally finite graph, and let $\mathcal{B}(n)$ be the ball of radius $n$ in $\Gamma$, based at some fixed vertex. Define $\|\Gamma \setminus \mathcal{B}(n)\|$ to be the number of connected, unbounded components in the complement of $\mathcal{B}(n)$.

**Lemma 11.18.** *Let $\Gamma$ be a locally finite graph and let $m < n$ be two positive integers. Then*

$$\|\Gamma \setminus \mathcal{B}(m)\| \le \|\Gamma \setminus \mathcal{B}(n)\|.$$

*Proof.* Let $\mathcal{C}$ be an unbounded, connected component of $\Gamma \setminus \mathcal{B}(m)$. Either $\mathcal{C}$ remains connected when $\mathcal{B}(n)$ is removed, or it does not. Thus each unbounded, connected component of $\Gamma \setminus \mathcal{B}(m)$ contributes at least one unbounded, connected component to $\Gamma \setminus \mathcal{B}(n)$. $\square$

**Definition 11.19** (Ends of a graph). Let $\Gamma$ be a connected, locally finite graph, and let $\mathcal{B}(n)$ be the ball of radius $n$ about a fixed vertex $v \in V(\Gamma)$. The *number of ends* of $\Gamma$ is

$$e(\Gamma) = \lim_{n \to \infty} \|\Gamma \setminus \mathcal{B}(n)\|.$$

This limit exists, although it may equal $\infty$, since, by Lemma 11.18, the sequence $\{\|\Gamma \setminus \mathcal{B}(n)\|\}$ is a non-decreasing sequence of integers.

**Example 11.20.** Let $\mathrm{Comb}_n$ be the graph formed by joining $n$ "rays" by a combinatorial path connecting their base vertices, as in Figure 11.4. Let $v$ be the vertex in the bottom left-hand corner. Then:

$$\|\mathrm{Comb}_n \setminus \mathcal{B}(0)\| = 2,$$
$$\|\mathrm{Comb}_n \setminus \mathcal{B}(1)\| = 3,$$
$$\|\mathrm{Comb}_n \setminus \mathcal{B}(2)\| = 4,$$

and so on, until $\|\mathrm{Comb}_n \setminus \mathcal{B}(n-1)\| = n$, after which

$$\|\mathrm{Comb}_n \setminus \mathcal{B}(k)\| = n$$

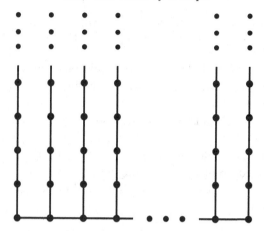

Fig. 11.4. The number of ends of "Comb$_n$" is $n$.

for all $k \geq n - 1$. Thus $e(\text{Comb}_n) = n$.

You may wonder why the adjective "unbounded" is included in the definition of $||\Gamma \setminus \mathcal{B}(n)||$. If $\Gamma \setminus \mathcal{B}(n)$ could contain a finite connected component, then such a component will eventually be contained in some $\mathcal{B}(m)$ for $m > n$. Thus that finite component is not a part of the graph that extends indefinitely. Further, such finite components can occur. At the end of the exercises in Chapter 9 there is a discussion of dead ends in Cayley graphs. This discussion and the following exercises show that if $\Gamma$ is the Cayley graph of the lamplighter group then $\Gamma \setminus \mathcal{B}(n)$ will have finite, connected components.

The definition of the number of ends of a graph $\Gamma$ uses a particular choice of a base vertex, but this vertex does not show up in the notation. This choice does not influence the value of $e(\Gamma)$ and, to show this, we extend an earlier definition. Let $C$ be *any* finite subgraph of $\Gamma$ and let $||\Gamma \setminus C||$ denote the number of unbounded, connected components of the complement of $C$. Let

$$e_c(\Gamma) = \text{Max}\{||\Gamma \setminus C|| \mid C \text{ is a finite subgraph of } \Gamma\}$$

be the largest value that occurs for all finite subgraphs $C$, or set $e_c(\Gamma) = \infty$ if there is no $n \in \mathbb{N}$ such that $n \geq ||\Gamma \setminus C||$ for all finite subgraphs $C \subset \Gamma$. (If you have studied real analysis, feel free to replace "Max" with "sup.")

**Lemma 11.21.** *Let* $\Gamma$ *be a connected, locally finite graph and let* $e(\Gamma)$ *and* $e_c(\Gamma)$ *be as defined above. Then* $e(\Gamma) = e_c(\Gamma)$.

*Proof.* Since metric balls $\mathcal{B}(n)$ are finite subgraphs in any locally finite graph $\Gamma$, $e(\Gamma) \leq e_c(\Gamma)$. Conversely, given any finite subgraph $C$, there is an $n \in \mathbb{N}$ such that $C \subset \mathcal{B}(n)$. Thus $||\Gamma \setminus C|| \leq ||\Gamma \setminus \mathcal{B}(n)||$, by the same argument used to prove Lemma 11.18, and therefore $e_c(\Gamma) \leq e(\Gamma)$. $\square$

It is natural to consider the number of ends of a Cayley graph of some finitely generated group $G$. In keeping with the theme of this chapter, we first establish that the resulting number is independent of the choice of Cayley graph.

**Lemma 11.22.** *Let* $S$ *and* $T$ *be two finite generating sets for* $G$, *and let* $\mathcal{B}_S(n)$ *and* $\mathcal{B}_T(n)$ *be the balls of radius* $n$ *in* $\Gamma_S$ *and* $\Gamma_T$, *respectively. Then there is a constant* $\mu \geq 1$ *such that if* $v_g$ *and* $v_h$ *are vertices in* $\Gamma_S$ *that can be joined by an edge path outside of* $\mathcal{B}_S(\mu n + \mu)$ *then* $v_g$ *and* $v_h$ *in* $\Gamma_T$ *are outside* $\mathcal{B}_T(n)$ *and can be joined by a path that stays outside of* $\mathcal{B}_T(n)$.

*Proof.* Let $\phi_{T \leftarrow S}$ be a map taking $\Gamma_S \rightarrow \Gamma_T$ as in Proposition 11.2. By Corollary 11.3, if $\lambda$ is at least the length of the longest expression chosen to represent the generators from $S$ in terms of words in $\{T \cup T^{-1}\}^*$ then

$$d_T(g, h) \geq \frac{1}{\lambda} d_S(g, h).$$

Set $\mu = \lambda^2 + 1$.

Let $\{v_g = v_0, v_1, v_2, \ldots, v_n = v_h\}$ be the vertices that occur on an edge path from $v_g$ to $v_h$ that is contained in $\Gamma_S \setminus \mathcal{B}_S(\mu n + \mu)$. Then

$$d_T(v_e, v_i) \geq \frac{1}{\lambda}(\mu n + \mu) > n + \lambda.$$

Thus each $v_i$ is outside of $\mathcal{B}_T(n + \lambda)$. If $e$ is the edge joining $v_i$ to $v_{i+1}$ in $\Gamma_S$ then the length of $\phi_{T \leftarrow S}(e)$ is at most $\lambda$. It follows that $\phi_{T \leftarrow S}(e)$ is outside of $\mathcal{B}_T(n)$. Thus $\phi_{T \leftarrow S}$ takes the path joining $v_g$ to $v_h$ in $\Gamma_S \setminus \mathcal{B}_S(\mu n + \mu)$ to a path connecting $v_g$ to $v_h$ in $\Gamma_T \setminus \mathcal{B}_T(n)$. $\square$

**Theorem 11.23.** *Let* $S$ *and* $T$ *be two finite generating sets for a group* $G$, *and let* $\Gamma_S$ *and* $\Gamma_T$ *be the corresponding Cayley graphs. Then*

$$e(\Gamma_S) = e(\Gamma_T).$$

*Proof.* By Lemma 11.22 if two vertices in $\Gamma_S$ can be connected in the complement of $\mathcal{B}_S(\mu n + \mu)$ then they can be connected in the complement of $\mathcal{B}_T(n)$. Thus $\|\Gamma_S \setminus \mathcal{B}_S(\mu n + \mu)\| \geq \|\Gamma_T \setminus \mathcal{B}_T(n)\|$. It follows that

$$\lim_{n \to \infty} \|\Gamma_S \setminus \mathcal{B}_S(n)\| \geq \lim_{n \to \infty} \|\Gamma_T \setminus \mathcal{B}_T(n)\|,$$

that is, $e(\Gamma_S) \geq e(\Gamma_T)$. The opposite inequality holds *mutatis mutandis*, hence we are done. □

As the number of ends of a Cayley graph for $G$ does not depend on the choice of Cayley graph, we may make the following definition.

**Definition 11.24.** Let $G$ be a finitely generated group. The *number of ends of $G$* is the number of ends of any of its Cayley graphs. By Theorem 11.23 this number is a property of the group $G$; it does not depend on the choice of Cayley graph. We denote the number of ends of a finitely generated group $G$ by $e(G)$.

**Exercise 11.25.** Show that a finitely generated group $G$ has zero ends if and only if $G$ is finite.

**Example 11.26.** Consider a finitely generated free group $\mathbb{F}_n$ where $n$ is at least 2. The Cayley graph, with respect to a basis, is a tree where every vertex has valence $2n$. Thus $\|\Gamma \setminus \mathcal{B}(k)\| = 2n(2n-1)^k$, and therefore $e(\mathbb{F}_n) = \infty$, for all $n \geq 2$.

## 11.5  The Freudenthal–Hopf Theorem

At the end of Section 11.4 we pointed out that finite groups have zero ends (Exercise 11.25) and we sketched an argument showing that free groups have infinitely many ends (Example 11.26). Here are two more examples that are easily verified: $\mathbb{Z}$ has two ends while $\mathbb{Z} \oplus \mathbb{Z}$ is one-ended. What other values can $e(G)$ take? This question was resolved in the early 1930s, and the answer is referred to as the *Freudenthal–Hopf Theorem*.

**Theorem 11.27** (Freudenthal–Hopf Theorem). *Every finitely generated group has either zero, one, two or infinitely many ends.*

*Proof.* Previous examples show that there are groups with zero, one, two or infinitely many ends. Thus we need to establish that there is no group $G$ where $e(G)$ is finite but $e(G) > 2$. Assume to the contrary that the Cayley graph $\Gamma$ of a finitely generated group $G$ has $k \geq 3$ ends. Note that this implies that $G$ must be an infinite group, and that there is a number $n \in \mathbb{N}$ such that $\Gamma \setminus \mathcal{B}(n)$ has $k$ unbounded, connected components.

Since $G$ is infinite there is an element $g \in G$ such that $d(e, g) > 2n$, where $v_g$ is in an unbounded, connected component of $\Gamma \setminus \mathcal{B}(n)$. Then $g \cdot \mathcal{B}(n) \cap \mathcal{B}(n) = \emptyset$. Further, since $g\mathcal{B}(n)$ is contained in an unbounded, connected component of $\Gamma \setminus \mathcal{B}(n)$, it divides this component into at least $k$ connected pieces, at least $(k - 1)$ of which are unbounded. Thus if $C = \mathcal{B}(n) \cup g \cdot \mathcal{B}(n)$ then $C$ is a finite subgraph of $\Gamma$ and $\|\Gamma \setminus C\| \geq 2k - 2$ (see Figure 11.5). It follows that $e_c(\Gamma)$ – as defined in Lemma 11.21 – is at least $2k - 2$. Thus $e(\Gamma) \geq 2k - 2 > k$, since $k \geq 3$. But this contradicts the claim that $e(G) = k$.                                                    □

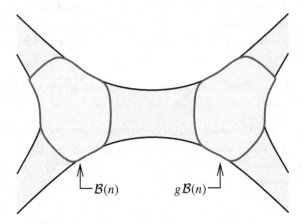

Fig. 11.5. If a group $G$ has at least three ends, then it has at least four ends, which means it has at least six ends, ...

**Exercise 11.28.** Copy and complete Table 11.1, filling in the values of $e(G)$. The formatting of this table is intended to help you guess the correct value of $e(G)$, even if you are not clear on how to actually prove that your guess is correct.

## 11.6 Two-Ended Groups

Given that every finitely generated group has zero, one, two or infinitely many ends, it is natural to ask which groups belong in a particular collection. We know that a group has zero ends if and only if it is finite. Looking over Table 11.1 it appears that the next simplest collection are the groups with two ends. The examples of such groups that come to

| $G$ | $e(G)$ |
|---|---|
| $\mathrm{Sym}_5$ | |
| $\mathbb{Z}_7 \rtimes \mathbb{Z}_3$ | |
| $(D_{397} \wr \mathbb{Z}_{736})^8$ | |

| $G$ | $e(G)$ |
|---|---|
| $\mathbb{Z} \oplus \mathbb{Z}$ | |
| $W_{244}$ | |
| $\mathrm{BS}(1,2)$ | |

| $G$ | $e(G)$ |
|---|---|
| $\mathbb{Z}$ | |
| $D_\infty$ | |
| $\mathbb{Z} \oplus \mathrm{Sym}_7$ | |

| $G$ | $e(G)$ |
|---|---|
| $\mathbb{F}_2$ | |
| $\mathbb{Z}_3 * \mathbb{Z}_4$ | |
| $(\mathbb{Z} \oplus \mathbb{Z}) * \mathbb{Z}_2$ | |

Table 11.1. *Exercise 11.28 asks you to fill in the values $e(G)$ in this table.*

mind are $\mathbb{Z}$, $D_\infty$, and any direct sum of a two-ended group with a finite group. In each case the group has a finite index subgroup $\approx \mathbb{Z}$, and the goal of this section is to establish that this is indeed always the case.

Recall that the *symmetric difference* of two sets is defined to be

$$A\triangle B = (A \cup B) \setminus (A \cap B).$$

(The symmetric difference was previously mentioned in our discussion of the lamplighter group in Chapter 8.) Since our discussion of two-ended groups will rely heavily on this operation, you should verify the following two formulas.

**Exercise 11.29.** Let $A, B$ and $C$ be sets. Then

$$A \triangle B = (A \triangle C) \triangle (C \triangle B).$$

Denote the complement of a set $X$ by $X^c$. Then

$$A \triangle B = A^c \triangle B^c.$$

For the remainder of this section, let $G$ be a finitely generated group with $e(G) = 2$, and let $\Gamma$ be a Cayley graph of $G$. There is then a finite subgraph $C$ such that $\Gamma \setminus C$ has exactly two connected, unbounded components. By adding all the finite components of $\Gamma \setminus C$ to $C$, we may assume that $\Gamma \setminus C$ consists of exactly two connected components, each of which is unbounded. Choose one of these two complements, and let $E \subset G$ consist of those elements of $G$ that correspond to the vertices in that component. Notice that we may apply elements of $G$ to $E$, forming subsets

$$gE = \{g \cdot h \mid h \in E\}.$$

Similarly we can left multiply $E^c$:

$$g \cdot E^c = \{g \cdot k \mid k \notin E\}.$$

Notice that $g \cdot E^c$ is the same set as $(g \cdot E)^c$. Thus we may simply write $gE^c$, avoiding the use of parentheses.

The proof of our first lemma is left as an exercise (Exercise 12 at the end of this chapter).

**Lemma 11.30.** *Let $G$ and $E$ be as above, and let $g \in G$. Then, because $G$ is two-ended, either $E \triangle gE$ is finite or $(E \triangle gE)^c$ is finite.*

**Lemma 11.31.** *Let $G$ and $E$ be as above. Then the subset of elements*

$$H = \{g \in G \mid E \triangle gE \text{ is finite}\}$$

*forms a subgroup of index at most 2.*

*Proof.* We first establish that $H$ is a subgroup. To show that it is closed under inverses, one only needs to note that

$$|E \triangle h^{-1}E| = |hE \triangle E|.$$

Hence if $E \triangle hE$ is finite, so is $E \triangle h^{-1}E$. To show that $H$ is closed under products, take two elements $g, h \in H$. Then

$$E \triangle ghE = (E \triangle gE) \triangle (gE \triangle ghE)$$
$$= (E \triangle gE) \triangle g (E \triangle hE).$$

Since we have assume $E \triangle gE$ and $E \triangle hE$ are both finite, so is $E \triangle ghE$.

In order to establish that the index of $H$ in $G$ is at most 2, assume that $H \neq G$ and that $g, h \in G \setminus H$. Then

$$
\begin{aligned}
E \triangle gh^{-1}E &= (E \triangle gE) \triangle (gE \triangle gh^{-1}E) \\
&= (E \triangle gE) \triangle g(E \triangle h^{-1}E) \\
&= (E \triangle gE)^{\mathrm{c}} \triangle g(E \triangle h^{-1}E)^{\mathrm{c}}.
\end{aligned}
$$

By Lemma 11.30, both $(E \triangle gE)^{\mathrm{c}}$ and $(E \triangle h^{-1}E)^{\mathrm{c}}$ are finite, hence so is $E \triangle gh^{-1}E$. □

Let $C$ be the finite subgraph that divided the Cayley graph $\Gamma$ into two unbounded, connected components, and let $g \in H$ be an element where $gC \cap C = \emptyset$. Then either $gC \subset E$ or $gC \subset E^{\mathrm{c}}$. From this and the definition of $H$, one gets:

**Lemma 11.32.** *Let $G$, $H$, $C$ and $E$ be as above, and let $g \in H$ be an element such that $C \cap gC = \emptyset$. Then either*

*1. $E \cap gE^{\mathrm{c}} = \emptyset$ and $E^{\mathrm{c}} \cap gE \neq \emptyset$, or*
*2. $E \cap gE^{\mathrm{c}} \neq \emptyset$ and $E^{\mathrm{c}} \cap gE = \emptyset$.*

We are now well-positioned to establish the main result of this section. Theorem 11.33 below was first established by C. T. C. Wall in 1967, using purely algebraic techniques. Our approach has been to build up an intuitive notion of the "right side" of the Cayley graph $\Gamma$, which is the part that contains $E$. On an intuitive level, an element $g \in H$ moves $\Gamma$ "to the right" if $gE \subset E$. Further, $g$ moves $\Gamma$ "far to the right" if $E \cap gE^{\mathrm{c}}$ is large. Surprisingly, this vague intuition can be made quite precise.

**Theorem 11.33.** *Let $G$ and $H$ be as above. Then there is a homomorphism $\phi : H \to \mathbb{Z}$ whose kernel is finite.*

*Proof.* Since $g \in H$, $E \triangle gE$ is finite, hence $E \cap gE^{\mathrm{c}}$ and $E^{\mathrm{c}} \cap gE$ are also finite. We may then define a function $\phi : H \to \mathbb{Z}$ by

$$
\phi(g) = |E \cap gE^{\mathrm{c}}| - |E^{\mathrm{c}} \cap gE|.
$$

Notice that $E \cap gE^{\mathrm{c}}$ is the disjoint union

$$
E \cap gE^{\mathrm{c}} = (E \cap gE^{\mathrm{c}} \cap ghE^{\mathrm{c}}) \cup (E \cap gE^{\mathrm{c}} \cap ghE)
$$

and similarly

$$
E^{\mathrm{c}} \cap gE = (E^{\mathrm{c}} \cap gE \cap ghE^{\mathrm{c}}) \cup (E^{\mathrm{c}} \cap gE \cap ghE).
$$

Thus we may also define $\phi(g)$ to be

$$\phi(g) = |E \cap gE^c \cap ghE^c| + |E \cap gE^c \cap ghE|$$
$$- |E^c \cap gE \cap ghE^c| - |E^c \cap gE \cap ghE|.$$

Now let $h \in H$. We have

$$\phi(h) = |E \cap hE^c| - |E^c \cap hE| = |gE \cap ghE^c| - |gE^c \cap ghE|.$$

By dividing $gE \cap ghE^c$ and $gE^c \cap ghE$ into the parts that are in $E$ and the parts that are in $E^c$, we have

$$\phi(h) = |E \cap gE \cap ghE^c| + |E^c \cap gE \cap ghE^c|$$
$$- |E \cap gE^c \cap ghE| - |E^c \cap gE^c \cap ghE|.$$

Cancelling terms and applying the same trick as above, but in reverse, we find

$$\phi(g) + \phi(h) = |E \cap gE^c \cap ghE^c| + |E \cap gE \cap ghE^c|$$
$$- |E^c \cap gE \cap ghE| - |E^c \cap gE^c \cap ghE|$$
$$= |E \cap ghE^c| - |E^c \cap ghE|$$
$$= \phi(gh).$$

Thus, not only is $\phi$ a function from $H$ to $\mathbb{Z}$, it is a group homomorphism.

We use Lemma 11.32 to see that the kernel of $\phi$ must be finite. As $C$ is a finite subgraph of a Cayley graph $\Gamma$, there are only finitely many elements where $C \cap gC \neq \emptyset$. But if $C \cap gC = \emptyset$ then Lemma 11.32 implies $\phi(g) \neq 0$. $\qquad\square$

**Corollary 11.34.** *A group $G$ is two-ended if and only if $G$ contains a finite index subgroup isomorphic to $\mathbb{Z}$.*

*Proof.* Theorem 11.33 states that $G$ has a finite index subgroup $H$ where there is a group homomorphism $\phi : H \to \mathbb{Z}$ with finite kernel. If $h \in H$ is any element where $\phi(h) \neq 0$, then $\langle h \rangle \approx \mathbb{Z}$ and $\langle h \rangle$ has finite index in $H$. Since either $G = H$ or $[G : H] = 2$ it follows that $\langle h \rangle$ has finite index in $G$.

Conversely, in Section 11.7 we show that if $H$ is a finite index subgroup of $G$, then $e(H) = e(G)$. In particular, since $e(\mathbb{Z}) = 2$, if $G$ has a subgroup $\approx \mathbb{Z}$, then $e(G) = 2$. $\qquad\square$

We now know that zero-ended groups are the same as finite groups, and two-ended groups are the same as groups that contain $\mathbb{Z}$ as a finite index subgroup. So in these two cases the connection between the structure of the group and the number of ends is clear. In 1968 John Stallings proved that given an infinitely ended group $G$, there is a tree $T$ and an action $G \curvearrowright T$. Further, this action satisfies a number of nice properties that allow you to express $G$ directly in terms of stabilizers of vertices and edges of $T$. A full description of this theorem is beyond the aims of this text, but there is a result that is simple to state.

**Theorem 11.35.** *Let $G$ be a finitely generated, torsion-free group. Then $G$ has infinitely many ends if and only if $G$ is a free product, $G \approx H * K$, where neither $H$ nor $K$ is the trivial group.*

The original proof of this result, under the stronger hypothesis that $G$ is finitely presented, can be found in [St68]. While some of the techniques used in Stallings' article are beyond a standard undergraduate experience, many parts of it are accessible. In fact, our discussion of two-ended groups is based on portions of this paper.

By repeatedly applying Theorem 11.35 to a finitely generated group, along with a few other powerful results, one eventually sees that every finitely generated, torsion-free group can be expressed as a free product of a free group with a finite number of one-ended groups (where the free group may be trivial or it might be the entire group). Since free groups are reasonably well understood, it is the one-ended groups that are mysterious. Here, indeed, there is great variety. Because of this, the development of more refined tools that can be used to explore what happens "at infinity" in an infinite group is of active interest. See for example [Ge08].

## 11.7 Commensurable Groups and Quasi-Isometry

When we introduced the Drawing Trick (Remark 1.49), we described it as a means of constructing a Cayley graph of a group. If $G \curvearrowright X$ where $X$ is some geometric object, and there is an $x \in X$ that $G$ moves freely, then there is a bijective correspondence between points in the orbit of $x$ and the elements of $G$. The set $\mathrm{Orb}(x)$ is something like a collection of polka dots decorating $X$, and often it provides an easy-to-visualize collection of vertices for a Cayley graph. Consider the elementary example of $\mathbb{Z} \oplus \mathbb{Z}$ acting on $\mathbb{R}^2$ with generators $(1,0)$ and $(0,1)$ being a horizontal translation and a vertical translation, respectively. The orbit of the origin under this action is the set of all points $(m,n)$ whose coordinates are integers. We refer to such points as *integral points*. We can view the collection of integral points as the vertices of a Cayley graph, $\Gamma$, embedded in $\mathbb{R}^2$, where the edges are arcs of length 1.

Given two integral points we can measure the distance between them in $\mathbb{R}^2$ using the standard metric:

$$d_{\mathbb{R}^2} = \sqrt{(m_1 - m_0)^2 + (n_1 - n_0)^2}.$$

We can also measure the distance between them in the Cayley graph of $\mathbb{Z} \oplus \mathbb{Z}$ or, in other words, in the word metric associated to the generators $\{(1,0), (0,1)\}$:

$$d_{\mathbb{Z}^2} = |m_1 - m_0| + |n_1 - n_0|.$$

How different are these two notions of distance? The distance between two points in the plane is realized by a straight line segment and, given any such line, there is an edge path in $\Gamma \subset \mathbb{R}^2$ that stays close to it (as in Figure 11.6). Because straight lines in $\mathbb{R}^2$ can be roughly approximated

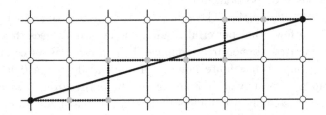

Fig. 11.6. Straight lines in the Euclidean plane can be roughly approximated by geodesic edge paths in an embedded Cayley graph for $\mathbb{Z} \oplus \mathbb{Z}$.

by edge paths in $\Gamma$, one expects that there is a way to move between the geometry of the Euclidean plane and the geometry of this Cayley graph. The straight line from $(0,0)$ to $(m, n)$ in $\mathbb{R}^2$ has length equal to $\sqrt{m^2 + n^2}$, which is less than (or occasionally equal to) the length of an edge path in $\Gamma$ between these points. On the other hand, we can always arrange it so that a straight line joining integral points is the hypotenuse of a right triangle, the sides of which sit in the Cayley graph $\Gamma$. Thus the distance in $\Gamma$ is bounded from above by twice the distance in $\mathbb{R}^2$ (or by $\sqrt{2}$ times the distance in $\mathbb{R}^2$ if you want to keep your constants minimal). If we let $\iota : \mathbb{Z}^2 \to \mathbb{R}^2$ be the polka-dot embedding of $\mathbb{Z} \oplus \mathbb{Z}$ in $\mathbb{R}^2$ that we have been discussing, then we have just established that for any two elements $x, y \in \mathbb{Z}^2$:

$$\frac{1}{2} d_{\mathbb{Z}^2}(x, y) \leq d_{\mathbb{R}^2}(\iota(x), \iota(y)) \leq 2 d_{\mathbb{Z}^2}(x, y)$$

where we have included an unnecessary "2" in the right-hand inequality in order to draw a parallel with the inequalities given in Corollary 11.3.

We are now in a situation that is similar to our consideration of different Cayley graphs of a fixed group. Both $\mathbb{R}^2$ and $\Gamma$ provide a geometric picture of $\mathbb{Z} \oplus \mathbb{Z}$, one using the Euclidean metric and the other using the word metric, and the distortion between them seems controllable. There is, however, one important difference that shows up when we construct a map $\mathbb{R}^2 \to \mathbb{Z}^2$. Define $\phi : \mathbb{R}^2 \to \mathbb{Z}^2$ to be a map sending $(x, y)$ to a closest integral point; when there is more than one closest integral point, such as when $(x, y) = (0.5, 0.5)$, simply choose one of the available options. Notice that $\phi$ will take some points which are close together and move them further apart. For example, $\phi[(0.4, 0.4)] = (0, 0)$ and $\phi[(0.6, 0.6)] = (1, 1)$. In fact, $\phi[(0.49 \cdots 99, 0.49 \cdots 99)] = (0, 0)$ and $\phi[(0.50 \cdots 01, 0.50 \cdots 01)] = (1, 1)$. Thus the map $\phi$ will take points that are arbitrarily close to each other in $\mathbb{R}^2$ to points that are a distance 2 apart in $\mathbb{Z}^2$. Hence there is no real number $\lambda \geq 1$ such that

$$d_{\mathbb{Z}^2}(\phi(x), \phi(y)) \leq \lambda d_{\mathbb{R}^2}(x, y)$$

for any $x, y \in \mathbb{R}^2$. What is true is that the distortion introduced by $\phi$ is also controllable, but one needs to allow for a small amount of local noise. Given any point $x \in \mathbb{R}^2$, the point $\phi(x)$ will be at most $1/\sqrt{2}$ away from $x$; we also know that $d_{\mathbb{R}^2}(\phi(x), \phi(y)) \leq d_{\mathbb{Z}^2}(\phi(x), \phi(y))$; combining this with the triangle inequality gives us

$$\begin{aligned} d_{\mathbb{R}^2}(x, y) &\leq d_{\mathbb{R}^2}(x, \phi(x)) + d_{\mathbb{R}^2}(\phi(x), \phi(y)) + d_{\mathbb{R}^2}(\phi(y), y) \\ &\leq d_{\mathbb{R}^2}(\phi(x), \phi(y)) + 2 \\ &\leq d_{\mathbb{Z}^2}(\phi(x), \phi(y)) + 2. \end{aligned}$$

We also know that $d_{\mathbb{Z}^2}(\phi(x), \phi(y)) \leq 2d_{\mathbb{R}^2}(\phi(x), \phi(y))$ as well as that $d_{\mathbb{R}^2}(\phi(x), \phi(y)) \leq d_{\mathbb{R}^2}(x, y) + \sqrt{2}$. So $d_{\mathbb{Z}^2}(\phi(x), \phi(y)) \leq 2d_{\mathbb{R}^2}(x, y) + 2\sqrt{2}$. Combining these results we get:

$$d_{\mathbb{R}^2}(x, y) - 2 \leq d_{\mathbb{Z}^2}(\phi(x), \phi(y)) \leq 2d_{\mathbb{R}^2}(x, y) + 2\sqrt{2}.$$

There is little reason to attempt to make these constants as small as possible, and we prefer a more symmetric statement than that given above, hence we conclude by noting that the inequalities above imply:

$$\tfrac{1}{3}d_{\mathbb{R}^2}(x, y) - 3 \leq d_{\mathbb{Z}^2}(\phi(x), \phi(y)) \leq 3d_{\mathbb{R}^2}(x, y) + 3.$$

**Definition 11.36.** Let $X$ and $Y$ be two metric spaces and let $\phi$ be a function from $X$ to $Y$. The function $\phi$ is a *quasi-isometric embedding* if there is a constant $\lambda \geq 1$ such that for all $x, x' \in X$:

$$\frac{1}{\lambda}d_X(x, x') - \lambda \leq d_Y(\phi(x), \phi(x')) \leq \lambda d_X(x, x') + \lambda.$$

(See Definition 11.7, where similar inequalities were used to define equivalent growth functions.)

The function $\phi$ is *quasi-dense*, which is also sometimes referred to as *quasi-onto*, if there is a $\lambda$ such that, for any point $y \in Y$, there is some $x \in X$ such that $d_Y(\phi(x), y) \leq \lambda$.

Finally, $\phi : X \to Y$ is a *quasi-isometry* if it is both a quasi-isometric embedding and is quasi-dense.

If $\phi$ is a quasi-isometry and $\lambda$ is large enough to satisfy all the requirements above, then $\lambda$ is referred to as the *quasi-isometry constant* of $\phi$.

There are a number of elementary examples of quasi-isometric spaces. Our discussion of the polka-dot embedding of $\mathbb{Z}^2$ in $\mathbb{R}^2$ shows that the group $\mathbb{Z}^2$ with its standard word metric is quasi-isometric to the Euclidean plane. The cyclic subgroup of $\mathbb{Z}^2$ generated by $(1,0)$ is quasi-isometrically embedded in $\mathbb{R}^2$, but it is not quasi-isometric to $\mathbb{R}^2$, as it is not quasi-dense. In addition to the specific case of $\mathbb{Z}^2 \curvearrowright \mathbb{R}^2$, we have also encountered a large collection of quasi-isometric metric spaces. If $G$ is a finitely generated group, and $S$ and $T$ are two finite generating sets, then the word metric on $G$ induced by $S$ and the word metric on $G$ induced by $T$ yield quasi-isometric metric spaces; this follows immediately from Corollary 11.3. As this is an important point, we record it as:

**Lemma 11.37.** *Let $G$ be a finitely generated group and let $S$ and $T$ be two finite generating sets. Then $G$ with the word metric induced by $S$ is quasi-isometric to $G$ with the word metric induced by $T$.*

Let $\phi : X \to Y$ be a quasi-isometry with associated constant $\lambda$. Define a function $\psi : Y \to X$ by $\psi(y) = x$, where $x \in X$ is chosen so that $d_Y(\phi(x), y) \leq \lambda$. (Such an $x$ always exists because $\phi$ is quasi-dense.) It follows from the definition of $\psi$ that

    1. $d_Y(\phi \circ \psi(y), y) \leq \lambda$ for all $y \in Y$.

Because $\phi$ is a quasi-isometry, with quasi-isometry constant $\lambda$, we know that
$$\frac{1}{\lambda}d_X(x, x') - \lambda \leq d_Y(\phi(x), \phi(x')) \leq \lambda d_X(x, x') + \lambda.$$

In particular, if $x' = \psi \circ \phi(x)$, then $\phi(x) = \phi(x')$, which implies that $d_Y(\phi(x), \phi(x')) = 0$. This establishes two additional facts:

    2. $d_X(x, \psi \circ \phi(x)) \leq \lambda$ for all $x \in X$.

3. Given any $x \in X$ there is a $y \in Y$ such that $d_X(x, \psi(y)) \leq \lambda$. (Take $y = \phi(x)$.)

We can rewrite the basic quasi-isometry inequalities for $\phi$ to read:

$$\frac{1}{\lambda}d_Y(\phi(x), \phi(x')) - \lambda \leq d_X(x, x') \leq \lambda d_Y(\phi(x), \phi(x')) + \lambda.$$

Thus, for any $y$ and $y' \in Y$,

$$\frac{1}{\lambda}d_Y(\phi \circ \psi(y), \phi \circ \psi(y')) - \lambda$$
$$\leq d_X(\psi(y), \psi(y'))$$
$$\leq \lambda d_Y(\phi \circ \psi(y), \phi \circ \psi(y')) + \lambda.$$

However, $d_Y(\phi \circ \psi(y), y) \leq \lambda$. So by the triangle inequality we have

$$d_Y(\phi \circ \psi(y), \phi \circ \psi(y'))$$
$$\leq d_Y(\phi \circ \psi(y), y) + d_Y(y, y') + d_Y(y', \phi \circ \psi(y'))$$
$$\leq d_Y(y, y') + 2\lambda.$$

By substitution we then get:

$$d_X(\psi(y), \psi(y')) \leq \lambda d_Y(y, y') + \left(2\lambda^2 + \lambda\right).$$

Combining this with a similar argument applied to the other inequality establishes:

4. There is a constant $\Lambda \geq 1$ such that, for any $y, y' \in Y$,

$$\frac{1}{\Lambda}d_Y(y, y') - \Lambda \leq d_X(\psi(y), \psi(y')) \leq \Lambda d_Y(y, y') + \Lambda.$$

Thus we have the following result.

**Lemma 11.38.** *If $\phi : X \to Y$ is a quasi-isometry between metric spaces, then there is a quasi-isometry $\psi : Y \to X$. Further, there is a constant $k$ such that, for all $x \in X$ and all $y \in Y$, $d_X(\psi \circ \phi(x), x) \leq k$ and $d_Y(\phi \circ \psi(y), y) \leq k$.*

If $X$ and $Y$ are metric spaces, we let $X \sim_{\text{QI}} Y$ denote the relation "$X$ is quasi-isometric to $Y$."

**Proposition 11.39.** *The relation $X \sim_{\text{QI}} Y$ is an equivalence relation.*

*Proof.* The identity map is a quasi-isometry from $X$ to $X$ hence $X \sim_{\text{QI}} X$ for all metric spaces $X$. If $X \sim_{\text{QI}} Y$, then Lemma 11.38 tells us there is a "quasi-inverse" that establishes $Y \sim_{\mathrm{QI}} X$. All that remains to be done is to establish the transitive property.

Let $\phi_1 : X \to Y$ and $\phi_2 : Y \to Z$ be two quasi-isometries. Let $\lambda$ be the maximum of the two associated quasi-isometry constants. The composition $\phi = \phi_2 \circ \phi_1$ is a function from $X$ to $Z$. Given any two points $x, x' \in X$ we know that

$$\frac{1}{\lambda} d_X(x, x') - \lambda \le d_Y(\phi_1(x), \phi_1(x')) \le \lambda d_X(x, x') + \lambda.$$

and that

$$\frac{1}{\lambda} d_Y(\phi_1(x), \phi_1(x')) - \lambda$$
$$\le d_Z(\phi_2 \circ \phi_1(x), \phi_2 \circ \phi_1(x'))$$
$$\le \lambda d_Y(\phi_1(x), \phi_1(x')) + \lambda.$$

Thus we have

$$\frac{1}{\lambda^2} d_X(x, x') - (\lambda + 1) \le d_Z(\phi(x), \phi(x')) \le \lambda^2 d_X(x), x') + (\lambda^2 + \lambda).$$

Setting $\Lambda = \lambda^2 + \lambda$, we find

$$\frac{1}{\Lambda} d_X(x, x') - \Lambda \le d_Z(\phi(x), \phi(x')) \le \Lambda d_X(x, x') + \Lambda,$$

so the function $\phi : X \to Z$ is a quasi-isometric embedding.

It remains to be shown that $\phi$ is quasi-dense. To do this, let $z$ be a point in $Z$. By hypothesis, there is some $y \in Y$ such that $d_Z(\phi_2(y), z) \le \lambda$. There is also an $x \in X$ such that $d_Y(\phi_1(x), y) \le \lambda$, and therefore

$$d_Z(\phi(x), \phi_2(y)) \le \lambda^2 + \lambda.$$

The triangle inequality implies

$$d_Z(\phi(x), z) \le d_Z(\phi(x), \phi_2(y)) + d_Z(\phi_2(y), z) \le \lambda^2 + 2\lambda.$$

Making our previously chosen value of $\Lambda$ a bit larger, $\Lambda = \lambda^2 + 2\lambda$, insures that the composition $\phi = \phi_2 \circ \phi_1$ is a quasi-isometry from $X$ to $Z$ with quasi-isometry constant $\Lambda$. Thus if $X \sim_{\mathrm{QI}} Y$ and $Y \sim_{\mathrm{QI}} Z$ then $X \sim_{\mathrm{QI}} Z$. □

The group $D_\infty$ acts on the real line, as was discussed in Chapter 2. Mimicking the discussion of $\mathbb{Z}^2 \curvearrowright \mathbb{R}^2$, it is not hard to establish that $D_\infty \sim_{\mathrm{QI}} \mathbb{R}$. If $a$ and $b$ are generating reflections, then $\tau = ab$ is a translation, and $\mathbb{Z} \approx \langle \tau \rangle < D_\infty$. Not surprisingly, $\mathbb{Z} \sim_{\mathrm{QI}} \mathbb{R}$. Because $\sim_{\mathrm{QI}}$ is an equivalence relation, it follows that $\mathbb{Z} \sim_{\mathrm{QI}} D_\infty$, which implies that a group can be quasi-isometric to one of its subgroups. In fact, this is always the case, assuming that the subgroup is of finite index. In order to establish this claim, we first need to prove the following result.

**Theorem 11.40.** *Let $H$ be a finite index subgroup of $G$. Then $H$ is finitely generated if and only if $G$ is finitely generated.*

*Proof.* First, assume that $G$ is finitely generated, and let $\Gamma$ be an associated Cayley graph. Thus $G \curvearrowright \Gamma$, where $\Gamma$ is a locally finite, connected graph. Further, a fundamental domain for this action is a finite subgraph. Since $H$ is a subgroup of $G$, $H$ also acts on $\Gamma$. A fundamental domain for the action of $H$ on $\Gamma$ consists of $[G : H]$ copies of a fundamental domain for $G \curvearrowright \Gamma$. Thus $H \curvearrowright \Gamma$ also has a finite fundamental domain, since $H$ has finite index in $G$. Using this fundamental domain for $H \curvearrowright \Gamma$ one can construct a finite generating set for $H$. (Use Theorems 1.55 and 1.58 to fill in the details of this argument.)

Conversely, assume $H$ has a finite generating set $\{h_1, \ldots, h_m\}$. Because $H$ has finite index in $G$, we can express $G$ as a union of finitely many $H$-cosets: $G = H \cup g_1 H \cup \cdots \cup g_n H$. Thus every $g \in G$ can be expressed as $g = g_i h$ for some $h \in H$, and therefore every $g \in G$ can be expressed as a word in $\{g_1, \ldots g_n, h_1, \ldots, h_m, h_1^{-1}, \ldots, h_m^{-1}\}^*$.  $\square$

Theorem 11.40 is useful in many contexts in the study of infinite groups. For us it is most useful in allowing us to view a finitely generated group $G$ *and* any finite index subgroup $H < G$ as metric spaces, where the distance function in each is given by a word metric. As any two word metrics for a given group yield quasi-isometric metric spaces, we rarely need to be specific about which word metric is being considered.

**Proposition 11.41.** *Let $H$ be a finite-index subgroup of $G$, both of which are finitely generated. Then $H \sim_{QI} G$.*

*Proof.* Let $\Gamma$ be the Cayley graph of $G$ with respect to a generating set $S$. Since $H$ is a subgroup of $G$ there is an action $H \curvearrowright \Gamma$. Since $H$ has finite index in $G$, there is a finite fundamental domain $\mathcal{F}$ for this action. Let $\mathcal{B}_R$ be the ball of radius $R$ in $\Gamma$, centered at the vertex corresponding to the identity, where $R$ is chosen large enough so that $\mathcal{F} \subset \mathcal{B}_R$. Since $H \cdot \mathcal{F} = \Gamma$, we also know $H \cdot \mathcal{B}_R = \Gamma$, hence the distance from any $g \in G$ to some $h \in G$ is at most $R$. So the embedding of the subgroup $H$ into $G$ is quasi-dense.

The same argument as is given for Theorem 1.55 shows that

$$T = \{h \in H \mid h\mathcal{B}_R \cap \mathcal{B}_R \neq \emptyset\}$$

is a generating set for $H$. Further, if $h \in T$ then the length of $h$, with respect to the generators $S \subset G$, is at most $2R$. Hence by replacing each $h \in T$ by a word of length $\leq 2R$ in $\{S \cup S^{-1}\}^*$ we prove that $d_H(h, h') \leq 2R d_G(h, h')$.

To establish the other inequality, notice that any finite edge path in $\Gamma$ is contained in a finite sequence $\{h_0 \mathcal{B}_R, h_1 \mathcal{B}_R, \ldots, h_k \mathcal{B}_R\}$ where $h_i \mathcal{B}_R \cap h_{i+1} \mathcal{B}_R \neq \emptyset$. Further, if the edge path contains $n$ edges, then $k \leq n$. Thus $d_H(h, h') \leq d_G(h, h')$. It follows that

$$\frac{1}{2R} d_H(h, h') - 2R \leq d_G(h, h') \leq 2R d_H(h, h') + 2R$$

and so the embedding of the subgroup $H$ into $G$ is a quasi-isometry. $\quad\square$

**Definition 11.42.** Two groups $G$ and $H$ are *commensurable* if there is a finite sequence $\{G = H_0, H_1, \ldots, H_n = H\}$ where either $H_i$ contains $H_{i+1}$ as a finite-index subgroup, or $H_i$ is a finite-index subgroup in $H_{i+1}$.

An induction argument, using Proposition 11.41, shows:

**Corollary 11.43.** *Let $G$ and $H$ be commensurable groups. Then $G$ is finitely generated if and only if $H$ is finitely generated. If they are finitely generated, then $G \sim_{QI} H$.*

**Example 11.44.** Consider the group $\mathbb{Z}_3 * \mathbb{Z}_4$ from Section 3.5 and $SL_2(\mathbb{Z})$. By Corollary 3.41, $\mathbb{Z}_3 * \mathbb{Z}_4$ has a finite-index subgroup isomorphic to $\mathbb{F}_6$. Proposition 3.7 shows that $SL_2(\mathbb{Z})$ has a finite-index subgroup isomorphic to $\mathbb{F}_2$. Proposition 3.12 shows that $\mathbb{F}_2$ contains a finite-index subgroup isomorphic to $\mathbb{F}_3$. A generalization of this is given as Exercise 15 in Chapter 3, which asked you to show that, for any $n \geq 2$, $\mathbb{F}_n$ is a finite-index subgroup in $\mathbb{F}_2$. Thus the sequence

$$\{\mathbb{Z}_3 * \mathbb{Z}_4, \mathbb{F}_6, \mathbb{F}_2, SL_2(\mathbb{Z})\}$$

satisfies the conditions of Definition 11.42, whence $\mathbb{Z}_3 * \mathbb{Z}_4 \sim_{QI} SL_2(\mathbb{Z})$.

A similar argument establishes a more general result, which we leave to the reader as Exercise 15.

**Remark 11.45.** While commensurable groups are quasi-isometric, the converse is not true. There are many quasi-isometric groups that are not commensurable. One way to construct such examples is via the Švark–Milnor Theorem, a general result that encompasses much of our discussion in this section. The theorem states that if $X$ is a "reasonable" metric space, and $G \curvearrowright X$ is "reasonable," then $G \sim_{QI} X$. To construct quasi-isometric groups that are not commensurable one then just needs to find "reasonable" actions of two groups on the same "reasonable" metric space, where the sizes of the associated fundamental domains are not rational multiples of each other. (An account of the Švark–Milnor Theorem and its consequences – including explanations of "reasonable" – can be found in [BrHa99].)

We have previously established that the growth functions for a given group $G$, with respect to two finite generating sets, are equivalent (Corollary 11.9) and that the number of ends of a group is not dependent on the choice of finite generating set (Theorem 11.23). These results also hold for quasi-isometric groups. For example, assume $\phi : H \to G$ is a quasi-isometry with associated constant $\lambda$. If $h \in H$ is in the ball of radius $n$ about the identity, then $\phi(h)$ must be in the ball of radius $\lambda n + \lambda$ about the identity (in $G$). Thus $\beta_H(n) \leq \beta_G(\lambda n + \lambda)$, so $\beta_H \preceq \beta_G$. The quasi-inverse $\psi : G \to H$ shows that $\beta_G \preceq \beta_H$, and therefore $\beta_G \sim \beta_H$. We leave the proof that the number of ends is a quasi-isometry invariant as an exercise (Exercise 16 at the end of this chapter).

**Theorem 11.46.** *Let $G$ and $H$ be finitely generated groups that are quasi-isometric. Then*

1. *if $\beta_G$ and $\beta_H$ denote growth functions of $G$ and $H$, then $\beta_G \sim \beta_H$;*
2. *$e(G) = e(H)$.*

Thus we know, for example, that $\mathbb{Z}^n$ is not quasi-isometric to $\mathbb{Z}^m$ (for $m \neq n$) because they have inequivalent growth rates. Similarly, $\mathbb{F}_2$ and the lamplighter group are not quasi-isometric because $e(\mathbb{F}_2) = \infty$ while $e(L_2) = 1$ (Exercise 9).

## Exercises

(1) The dihedral group $D_3$ can be generated by two reflections or a reflection and a rotation, and the associated Cayley graphs are shown in Figure 11.1. Construct a map from the Cayley graph on the right to the Cayley graph on the left in this figure, along the lines discussed in the proof of Proposition 11.2.

(2) Let $\preceq$ be the relation on growth functions defined by $f \preceq g$ if there is a constant $\lambda \geq 1$ such that

$$f(x) \leq \lambda g(\lambda x + \lambda) + \lambda$$

for all $x \in [0, \infty)$. Show that $\preceq$ is symmetric and transitive, and that the relation $f \sim g$ – defined to mean both $f \preceq g$ and $g \preceq f$ is an equivalence relation

(3) Prove that if $c < d$, with $c$ and $d \in \mathbb{N}$, then $n^c \preceq n^d$ but $n^d \npreceq n^c$.

(4) Show that $2^{\sqrt{n}}$ is a growth function and that

$$n^d \prec 2^{\sqrt{n}} \prec 2^n$$

for any $d \in \mathbb{N}$. (In other words, show that $2^{\sqrt{n}}$ is an intermediate growth function.)

(5) Let $G$ and $H$ be finitely generated groups of polynomial growth. Show that $G \oplus H$ also has polynomial growth.

(6) Let $H$ be a subgroup of $G$, both finitely generated. Let $\beta_H$ and $\beta_G$ be growth functions for these groups. Show that $\beta_H \preceq \beta_G$.

(7) Prove that a finitely generated group $G$ has linear growth if and only if it is virtually cyclic.

(8) Prove that if $H$ is a finite index subgroup of $G$ then $e(H) = e(G)$.

(9) Prove that the lamplighter group is one-ended.

(10) Show that $e(\mathbb{F}_2 \oplus \mathbb{F}_2) = 1$.

(11) Let $G$ and $H$ be finitely generated, infinite groups. Prove that $e(G \oplus H) = 1$. (Warning: The existence of "dead ends" in Cayley graphs indicates that some care needs to be taken in establishing this claim.)

(12) Prove Lemma 11.30.

(13) Prove that $\langle b \rangle$ is not a quasi-isometric embedding of $\mathbb{Z}$ into $\mathrm{BS}(1, 2)$.

(14) Prove that the Euclidean reflection groups $W_{244}, W_{236}$ and $W_{333}$ are commensurable with each other.

(15) Let $G = G_1 * G_2$ and $H = H_1 * H_2$ be two free products of nontrivial finite groups, where at least one $G_i$ and at least one $H_i$ has order greater than 2. Prove $G \sim_{\mathrm{QI}} H$.

(16) Prove that the number of ends is a quasi-isometry invariant. That is, if $G$ and $H$ are quasi-isometric groups, then $e(G) = e(H)$.

# Bibliography

[An71] A.V. Anīsīmov, The group languages, *Kibernetika (Kiev)* **4** (1971) 18–24.

[BBN59] G. Baumslag, W. W. Boone, and B. H. Neumann, Some unsolvable problems about elements and subgroups of groups, *Math. Scand.* **7** (1959) 191–201.

[BS62] G. Baumslag and D. Solitar, Some two-generator one-relator non-Hopfian groups, *Bull. Am. Math. Soc.* **68** (1962) 199–201.

[BeBr05] J. Belk and K. S. Brown, Forest diagrams for elements of Thompson's group $F$, *Int. J. Algebra Comput.* **15** (2005) 815–50.

[BeBu05] J. Belk and K-U. Bux, Thompson's group $F$ is maximally nonconvex, in *Geometric Methods in Group Theory*, Cont. Math. **372** (2005) 131–46.

[Be97] C. Bennett, Explicit free subgroups of Aut($\mathbb{R}, \leq$), *Proc. Am. Math. Soc.* **125** (1997) 1305–1308.

[BjBr05] A. Björner and F. Brenti, *Combinatorics of Coxeter Groups*, Graduate Texts in Mathematics **231**, Springer, 2005.

[Bo87] O. V. Bogopolski, Arboreal decomposability of the group of automorphisms of a free group, *Algebra i Logica* **26** (1987) 131–49.

[BRS07] N. Brady, T. Riley, and H. Short, *The Geometry of the Word Problem for Finitely Generated Groups*, Birkhäuser, 2007.

[Bz94] M. Brazil, Growth functions for some nonautomatic Baumslag-Solitar groups, *Trans. Am. Math. Soc.* **342** (1994) 137–54.

[Br08] M. Bridson, Conditions that prevent groups from acting non-trivially on trees, *Geometry & Topology Monographs* **14**, Geometry & Topology Publications, (2008) 129–133.

[BrHa99] M. Bridson and A. Haefliger, *Metric Spaces of Non-Positive Curvature*, Springer, 1999.

[Ca84] J. W. Cannon, The combinatorial structure of cocompact discrete hyperbolic groups, *Geom. Dedicata* **16** (1984) 123–48.

[Ca87] J. W. Cannon, Almost convex groups, *Geom. Dedicata* **22** (1987) 197–210.

[CFP96] J. W. Cannon, W. J. Floyd, and W. R. Parry, Introductory notes on Richard Thompson's groups, *Enseign. Math. (2)* **42** (1996) 215–56.

[Ca78] A. Cayley, The theory of groups: Graphical representation, Desiderata and Suggestions No. 2 *Am. J. Math.* **1** (1878) 174–6.

[CT05] S. Cleary and J. Taback, Dead end words in lamplighter groups and other wreath products, *Quarterly J. Math.* **56** (2005) 165–78.

[CET06] S. Cleary, M. Elder, and J. Taback, Cone types and geodesic languages for lamplighter groups and Thompson's group $F$, *J. Algebra* **303** (2006) 476–500.

[CEG94] D. J. Collins, M. Edjvet, and C. P. Gill, Growth series for the group $\langle x, y | x^{-1}yx = y^l \rangle$, *Arch. Math. (Basel)* **62** (1994) 1–11.

[CL90] J. H. Conway and J. C. Lagarias, Tiling with polyminoes and combinatorial group theory, *J. Comb. Theory, A* **53** (1990) 183–208.

[CV96] M. Culler and K. Vogtmann, A group-theoretic criterion for property FA. *Proc. Am. Math. Soc.* **124** (1996) 677–83.

[Da08] M. W. Davis, *The Geometry and Topology of Coxeter Groups*, London Mathematical Society Monographs, Princeton University Press, 2008.

[De12] M. Dehn, Über unendliche diskontinuierliche gruppen, *Math. Ann.* **71** (1912) 116–44.

[Di94] W. Dicks, Equivalence of the strengthened Hanna Neumann conjecture and the amalgamated graph conjecture, *Invent. Math.* **117** (1994) 373–89.

[EH05] M. Elder and S. Hermiller, Minimal almost convexity, *J. Group Theory* **8** (2005) 239–66.

[E+92] D. Epstein, J. W. Cannon, D. Holt, S. V. F. Levy, M. S. Paterson, and W. P. Thurston, *Word Processing in Groups*, Jones and Bartlett, Boston, 1992.

[Ge08] R. Geoghegan, *Topological Methods in Group Theory*, Springer, 2008.

[Gi87] R. Gilman, Groups with a rational cross-section, in S. M. Gersten and J. R. Stallings (eds), *Combinatorial Group Theory and Topology*, Princeton University Press, 1987, pp. 175–83.

[Gi05] R. Gilman, Formal languages and their application to combinatorial group theory, in *Groups, Languages, Algorithms*, Cont. Math. **378**, American Mathematical Society, 2005, pp. 1–36.

[Gl92] A. M. W. Glass, The ubiquity of free groups, *Math. Intelligencer* **14** (1992) 54–7.

[Gr80] R. I. Grigorchuk, On Burside's problem on periodic groups, *Funktsional. Anal. i Prilozhen* **14** (1980) 53–4.

[GrPa07] R. I. Grigorchuk and I. Pak, Groups of intermediate growth: an introduction for beginners, to appear, *L'Enseignement Mathématique*.

[Gr81] M. Gromov, Groups of polynomial growth and expanding maps, *Publ. math. de l' I.H.E.S.*, **53** (1981) 53–78.

[Gr87] M. Gromov, Hyperbolic groups, in S. M. Gersten (ed.), *Essays in Group Theory*, Math. Sci. Res. Inst. Publ. 8, Springer, 1987, pp. 75–263.

[GrMa64] I. Grossman and W. Magnus, *Groups and their Graphs*, Mathematical Association of America, 1964.

[Gv96] J. R. J. Groves, Mimimal length normal forms for some soluble groups, *J. Pure Appl. Algebra* **114** (1996) 51–8.

[Gu04] V. S. Guba, On the properties of the Cayley graph of Richard Thompson's group $F$, *Int. J. Algebra Comput.* **14** (2004) 677–702.

[GuSa97] V. S. Guba and M. V. Sapir, The Dehn function and a regular set of normal forms for R. Thompson's group $F$, *J. Austral. Math. Soc. Ser. A* **62** (1997) 315–28.

[GuSi83] N. Gupta and S. Sidki, On the Burside problem for periodic groups, *Math. Z.* **182** (1983) 385–8.

[Hi51] G. Higman, A finitely related group with an isomorphic proper factor group, *J. Lond. Math. Soc.* **26** (1951) 59–61.

[Ho81] D. F. Holt, A graph which is edge transitive but not arc transitive, *J. Graph Theory* **5** (1981) 201–4.

[HoMe04] P. Hotchkiss and J. Meier, The growth of trees, *The College Mathematics Journal* **35** (2004) 143–51.

[Hu90] J. E. Humphreys, *Reflection Groups and Coxeter Groups*, Cambridge Studies in Advanced Mathematics **29**, Cambridge University Press, 1990.

[Jo97] D. L. Johnson, *Presentations of Groups*, 2nd edition, Cambridge University Press, 1997.

[Ku33] A. Kurosch, Über freie Produkte von Gruppen, *Math. Ann.* **108** (1933) 26–36.

[MKS66] W. Magnus, A. Karrass, and D. Solitar, *Combinatorial Group Theory*, John Wiley, 1966. (Reprinted by Dover)

[Me72] S. Meskin, Non-residually finite one-relator groups, *Trans. Am. Math. Soc.* **64** (1972) 105–14.

[Mi92] C. F. Miller, Decision problems for groups – survey and reflections, in G. Baumslag and C. F. Miller (eds), *Algorithms and Classification in Combinatorial Group Theory*, Math. Sci. Res. Inst. Publ. **23**, Springer, 1992 pp. 1–59.

[MiS98] C. F. Miller and M. Shapiro, Solvable Baumslag-Solitar groups are not almost convex, *Geom. Dedicata* **72** (1998) 123–7.

[MuS83] D. Muller and P. Schupp, Groups, the theory of ends and context-free languages, *J. Computer and System Sciences* **26** (1983) 295–310.

[Ne90] W. D. Neumann, On intersections of finitely generated subgroups of free groups, in L. G. Kovacs (ed.), *Groups: Canberra 1989*, Springer, 1990, pp. 161–70.

[Ni24] J. Nielsen, Die Isomorphismengruppe der freien Gruppen, *Math. Ann.* **91** 169–209.

[Ol95] A.Yu. Ol'shanskii, A simplification of Golod's example, in A. C. Kim and D. L. Johnson (eds), *Groups: Korea '94 (Pusan)*, de Gruyter, 1995, pp. 263–5.

[Pa92] W. Parry, Growth series of some wreath products, *Trans. Am. Math. Soc.* **331** (1992) 751–9.

[Re98] S. Rees, Hairdressing in groups: a survey of combings and formal languages, in *The Epstein Birthday Shrift*, Geometry and Topology Monographs **1**, 1998, pp. 493–509.

[Se77] J-P. Serre, *Arbres, Amalgams, SL₂*, Astérisque No. 46, Société Mathématique de France, 1977.

[Se80] J-P. Serre *Trees* (trans. John Stilwell), Springer-Verlag, 1980.

[St68] J. Stallings, On torsion-free groups with infinitely many ends, *Ann. Math.* **88** (1968) 312–34.

[St96] M. Stoll, Rational and transcendental growth series for the higher Heisenberg groups, *Invent. Math.* **126** (1996) 85–109.

[Th90] W. P. Thurston, Conway's tiling Groups, *Am. Math. Monthly* **97** (1990) 757–73.

[Ti72] J. Tits, Free subgroups in linear groups, *J. Algebra* **20** (1972) 250–270.

[Wa86] S. Wagon, *The Banach-Tarski Paradox*, Cambridge University Press, 1986.

[Wh88] S. White, The group generated by $x \mapsto x + 1$ and $x \mapsto x^p$ is free, *J. Algebra* **118** (1988) 408–22.

[Wi06] H. S. Wilf, *Generating Functionology*, 3rd edition, A. K. Peters, 2006.

# Index

Printed in the United States
by Baker & Taylor Publisher Services